Secrets
of Creation

Volume 3

Prime Numbers, Quantum Physics
and a Journey to the Centre of Your Mind

Secrets
of Creation

Volume 3

Prime Numbers, Quantum Physics
and a Journey to the Centre of Your Mind

Matthew Watkins

Winchester, UK
Washington, USA

First published by Liberalis Books, 2015
Originally published by The Inamorata Press, 2013
Liberalis Books is an imprint of John Hunt Publishing Ltd., Laurel House, Station Approach,
Alresford, Hants, SO24 9JH, UK
office1@jhpbooks.net
www.liberalisbooks.com

For distributor details and how to order please visit the 'Ordering' section on our website.

ISBN: 978 1 78279 777 7
Library of Congress Control Number: 2014958368

A CIP catalogue record for this book is available from the British Library.

The frontispiece is adapted from Matthäus Merian's
late 17th century engraving *Tabula Smaragdina*.

www.secretsofcreation.com

Printed and bound by CPI Group (UK) Ltd, Croydon, CR0 4YY, UK

We operate a distinctive and ethical publishing philosophy in all
areas of our business, from our global network of authors to
production and worldwide distribution.

table of contents

We have all this evidence that the Riemann zeros are vibrations,
but we don't know what's doing the vibrating.

Marcus du Sautoy, 2003

For Lilli

Prime Numbers, Quantum Physics and a Journey to the Centre of Your Mind is the final volume of the *Secrets of Creation* trilogy. It follows Volume 1: *The Mystery of the Prime Numbers* and Volume 2: *The Enigma of the Spiral Waves*.

This volume is written with the assumption that you've recently finished reading Volume 2, or at least that the ideas which it contains are relatively fresh in your mind. If they aren't, then I'd strongly recommend having a copy of the first two volumes available to consult as you read this one.

The notes contain a number of website addresses. If you find any of these to have become inaccessible, the Internet Archive's "Wayback Machine" at www.archive.org is a useful tool for recovering old versions of webpages.

More information on the *Secrets of Creation* trilogy, further web links, additional resources and an ever-expanding list of thanks and acknowledgements can be found at www.secretsofcreation.com.

Matthew Watkins
Canterbury, 2013

a ré-réintroduction

So where were we?

In the first book we had a look at prime numbers, particularly the way they distribute among the sequence of counting numbers, and we found that this distribution can be understood as being made up of an infinite sequence of "spiral waves".

The frequencies of these were left as a mystery to be explored in the second book, in which we considered the Riemann zeta function, particularly its "nontrivial zeros". The distances of these from the real axis (14.134b, 21.022…, 25.010…, …) give us the frequencies of the spiral waves. These zeros are known to be infinite in number and to all lie in a narrow strip, seemingly all on the central line of that strip. The question of whether or not they *do* all lie on that line, the subject of the Riemann Hypothesis, is seen as of the utmost importance for the structure of the number system. This, it was explained, is because the distances of the Riemann zeta zeros from the imaginary axis give us the *amplitude growth rates* of the spiral waves. If these aren't all the same (that is, 1/2), a kind of "balance" within the number system would be upset.

> "[T]he Riemann Hypothesis can be poetically (but rather accurately) reformulated as stating that ℚ, the field of rational numbers, lies as harmoniously as possible within the field of real numbers, ℝ. Since the ring of integers ℤ – and hence its field of fractions, ℚ – is arguably the most basic and fundamental object of all of mathematics, because it is the natural receptacle for elementary arithmetic, one may easily understand the centrality of the Riemann Hypothesis in mathematics…" (Michel Lapidus[1])

Volume 2 ended with a brief excursion into ideas about *spectral geometry* and spectra of various kinds, produced by various phenomena, made up of *eigenvalues*. This was all laid out for you so that I could drop the second cliffhanger ending: the extraordinary fact that *the "heights" of the zeta zeros appear to be the spectrum of eigenvalues of an unknown system*, something akin to an oscillating physical system.

This is the so-called *spectral interpretation* and it's backed up by an overwhelming body of evidence. *There's something "vibrating behind the number system" and we don't know what it is* (but we do have some clues). That, and its mind-twisting implications, are what this final volume will attempt to address.

I can happily confess that it's the content of this volume which provided the dominant motivation for me to write this trilogy. The mathematical (that is, verifiable, non-speculative) content of those first two books was just a necessary precursor. In order to appreciate the true depths of weirdness involved in what you're about to read, you're going to need the mental tools you acquired (or perhaps just polished) in reading Volumes 1 and 2.

chapter 28

the spectral interpretation: Exhibit A

It's now almost universally accepted that the zeta zero heights are a spectrum of some kind. As purely mathematical entities (unlike the light of a star or the warble of a bird), for them to be a spectrum means that it would have to be the spectrum of eigenvalues associated with a purely mathematical object known as an *operator*. The details of this operator, assuming it exists, remain a huge mathematical mystery.

The idea that the nontrivial zeta zeros might correspond to some kind of spectrum first appeared in the early 20th century. It's attributed to both David Hilbert and George Pólya who independently suggested an approach to proving the Riemann Hypothesis which involved interpreting the zeta zeros as the spectrum of a particular type of operator known as a *Hermitian operator* [1]. This suggestion could almost be classed as wishful thinking – particularly inspired wishful thinking, though.

The kind of Hermitian operator they had in mind would be associated with a kind of mathematical structure Hilbert had first described, now known as a *Hilbert space*. Hilbert spaces were employed to great effect in the early 20th century in the new branch of physics which became known as *quantum mechanics*.

Despite his meticulous research for the writing of his book *Prime Obsession*, John Derbyshire claims he could find no material evidence linking David Hilbert to this idea, so this must be accepted as part of "mathematical folklore". Assuming Hilbert *did* have this idea, we have no idea what brought him to it [2].

George Pólya, in correspondence with Andrew Odlyzko, provided a written account of his involvement with the idea (this was in 1982, some seventy years later). He explained that sometime around 1912–14, he was in the German university town of Göttingen, studying with the number theorist Edmund Landau. Landau had been very much aware of the RH and its significance for some years. In fact, it was his book on analytic number theory in 1909 which did more than anything else to bring this matter to the awareness of the wider mathematical community. One day, Pólya remembered, Landau asked him: "*You know some physics. Do you know a physical reason that the Riemann hypothesis should be true?*"[3]

Although now remembered as a pure mathematician, Pólya was aware at that time of the early work that had been done on quantum mechanics and the mathematical techniques which were involved. Using that kind of thinking, he suggested that if the nontrivial zeros could be interpreted in terms of the spectrum of a certain type of Hermitian operator, then (because of the specific properties of Hermitian operators) they would all be forced to lie on the critical line. Pólya never published this remark, but, in his words, "*…somehow it became known and it is still remembered.*"[4]

What, if anything, Hilbert was thinking will probably remain a mystery. In any case, this has become known as the "Hilbert–Pólya idea". The idea, remember, is that there is an operator with a spectrum which gives the "heights" of the Riemann zeta zeros. This operator, which is widely believed to exist, but which no one has yet been able to find, is sometimes referred to as the *Hilbert–Pólya operator*, sometimes as the *Riemann operator*.

There's now some circumstantial evidence that Riemann himself may have been thinking along these lines. In his time, mathematics and physics were not such clearly distinguished subjects. It was quite natural for a mathematician to spend time working on physics problems, the notion of "pure mathematics" not having fully crystallised in the way that it now has. It was recently discovered that at the same time Riemann was carrying out his research on the zeta function, he was also working on a physics problem involving a rotating ball of gas. This rotating ball of gas is a dynamical system and it has a spectrum – a spectrum of what are called *perturbation frequencies*. This connection was first noticed around the Millennium by Jonathan Keating when looking at Riemann's notes in the library at Göttingen University where he had worked. There's an account of this in Derbyshire's book, the author commenting "*...it's hard to believe that a mind as acute and penetrating as Riemann's would have missed the analogy between the zeta zeros strung out on the critical line, and his spectrum of perturbation frequencies...*"[5]

But Riemann never published anything about this which could have influenced Hilbert or Pólya (directly or indirectly). They were influenced by the physics of the time – quantum mechanics, with its use of complex numbers, operators and matrices. Spectra of eigenvalues as points strung out in lines on the complex plane would have been very much in the mathematical air, at least for those mathematicians such as Hilbert and Pólya who were keeping up with the contemporary physics. I should stress that neither Hilbert nor Pólya was suggesting that there's any "physicality" to the zeta zeros. They were simply applying some mathematical devices which would normally be relevant in the study of physical systems.

After decades as a vague mathematical possibility, evidence began to emerge in support of the Hilbert–Pólya approach. Consequently, what's become known as the spectral interpretation of the zeta zeros has now achieved near universal acceptance.

THE SELBERG TRACE FORMULA

There are two lines of evidence to support the spectral interpretation. They involve very different branches of mathematics, but they've come to be interrelated as a result of some curious new physics which itself has become associated with the interpretation.

The first evidence came in 1956 when the prolific Norwegian mathematician Atle Selberg published a formula in an Indian mathematics journal[6], now known as the *Selberg Trace Formula*, a mathematical breakthrough of the highest order (I'll abbreviate this as "STF"):

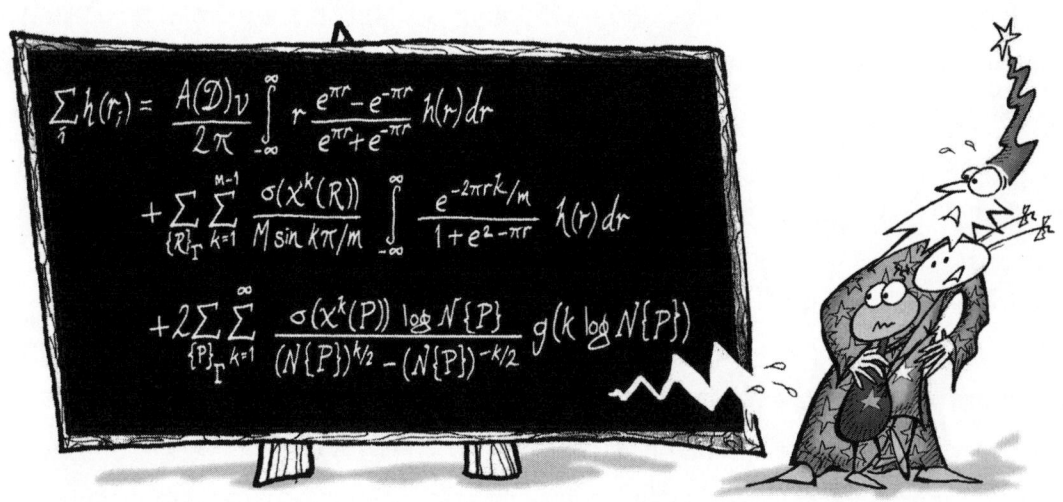

To the vast majority of people, it looks utterly alien, or even terrifying (unlike the relatively few people who are intimately familiar with it, to whom it's a thing of great beauty [7]). But, despite what most people concerned might think, it is possible to explain the basic idea of this formula to non-mathematicians.

What this formula is "about" is a family of geometrical objects called *compact Riemann surfaces*. It's Bernhard Riemann again, although his work on these "surfaces" was entirely separate from his brief yet deeply fertile foray into number theory in the late 1850s. We can only wonder if he would have been surprised had he been somehow shown that a century or so later, Riemann surfaces were to have become involved in the quest to understand his zeta function.

COMPACT RIEMANN SURFACES

To gain a real understanding of what a compact Riemann surface is would require an education in higher mathematics, but all we need to know for now is that it's a geometric object, in the same sort of way that a plane, a sphere, a cylinder or a torus (donut shape) is a geometric object. Unlike those examples (but in common with many other geometric objects mathematicians deal with) *it's not possible to construct Riemann surfaces in ordinary three-dimensional space.* Rooted as we all are in 3-d space, some people find it hard to accept that such things "exist". But we're able to examine, measure, test theories on, and otherwise study these things in seemingly endless detail. And we can all agree on everything that's discovered. So even if you're only prepared to accept them as a "shared fiction", keep in mind that they have a much stronger claim to "reality" than anything else we might normally think of as fiction.

Despite the impossibility of accurate three-dimensional models, it is possible to get a feeling for Riemann surfaces courtesy of the artist M.C. Escher, a master of making "impossible" things visible. First consider these images, inspired by ones he created[8]:

What we see here are a couple of *tessellations* (tilings) of what's known as the *hyperbolic disc*. There are certain repeating shapes which can be used to tile the familiar two-dimensional plane – squares, triangles, hexagons...

...or less regular shapes created by deforming these...

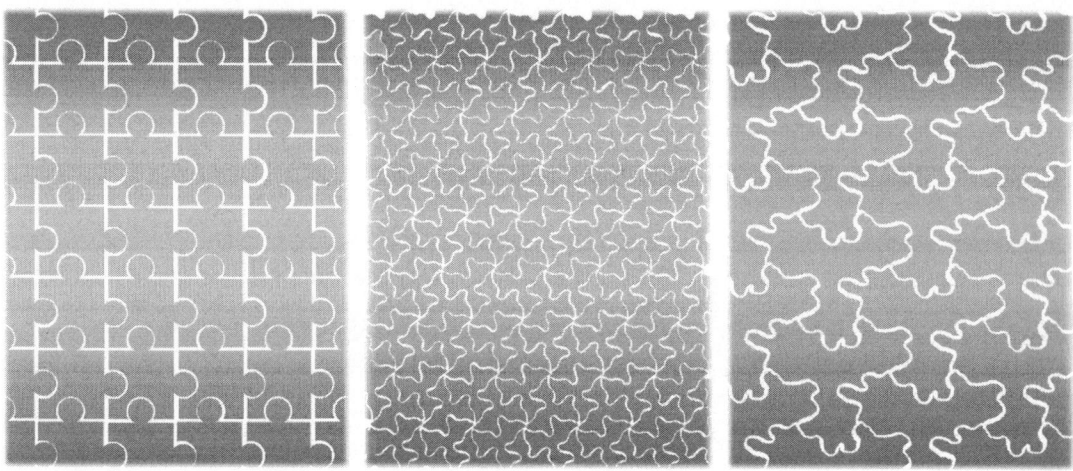

...but, in each case, all the "tiles" are the same size and shape. In the Escher-like images, we have a disc being tiled by repeating shapes which are of differing sizes, and curved into slightly different shapes. Most noticeably, they get smaller the closer they are to the edge.

To someone familiar with hyperbolic geometry, however, those wizards, *etc.* are *all the same size*. The measurement of space in the strange world of the hyperbolic disc is not like the measurement of space that you're used to. But it's possible to give a mathematically consistent definition of "distance" (and "angle") which can be applied on the disc in such a way that, when you measure them, each of the wizards and apprentices is found to be the same size (and shape).

There are a couple of ways of making sense of this. One is to pretend that you're not looking at a disc, but rather at an infinite curved surface "coming out at you" in 3-d, so the middle is the closest to you, and the closer to the edge you look on the disc, the "farther away" those points are (hence the wizards, *etc.* there look smaller).

Another way of looking at this involves the inhabitants of an imaginary disc-shaped world where the closer you get to the edge, the colder it becomes. Suppose there's a precise mathematical relationship between the distance from the edge and the temperature where you are. As the inhabitants, carrying rulers, approach the edge of their world, they *and* their rulers both shrink due to the cold, so their sense of distance is (from our point of view) distorted, but *there's no reason why they should ever notice this*[9]. If the distance-temperature relationship is set up properly, the inhabitants would then conclude that the peculiar wizard-shaped markings which covered their world were *all of identical size and shape*.

9

Escher used a combination of creative imagination and pure geometry[10] to bring forth his visions of "hyperbolically tessellating" shapes. If we put aside Escher's imagination and just consider the geometry behind it, we can see that there are certain geometric tiling patterns which subdivide the disc into regions which are, "hyperbolically" speaking, all the same size and shape.

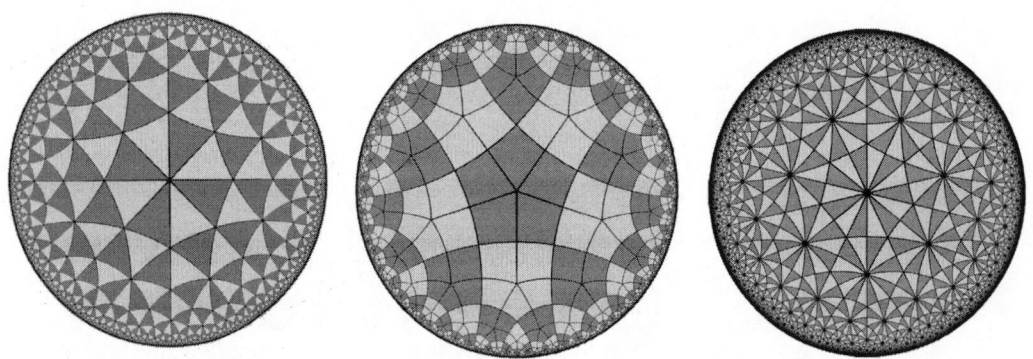

Some of these tilings of the hyperbolic disc are of interest to number theorists as they can be shown to be related (somewhat surprisingly) to number theory.

Geometrically, it's possible to think of (although not so easy to visualise) the disc suddenly "curling up" in such a way that all of the copies of a particular tile end up stacking neatly on top of each other. What we then get is an object which is effectively a single tile with its edges stitched together in a particular way.

The closest we can come to visualising this is by returning to our normal (Euclidean) geometry and considering the plane, tiled with squares to produce the familiar grid pattern.

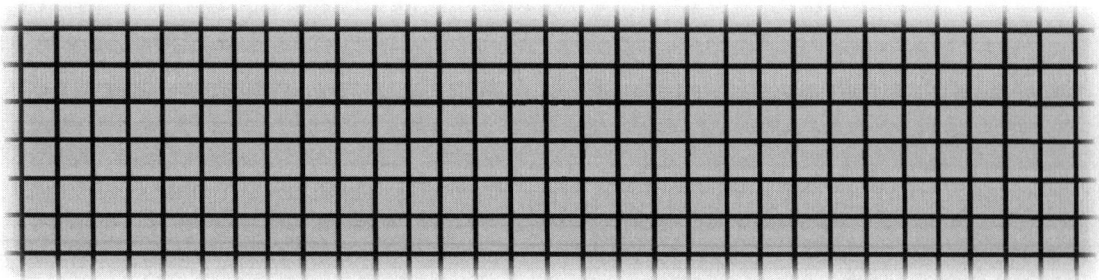

Now imagine that we were to roll the (infinitely thin) plane up into a tube so that all squares with the same "north-south position" end up stacked on top of each other:

Now we want these stacks of squares (each of which is now a cylindrical segment of the tube) to end up stacked on top of each other. This can be achieved, with some imaginative effort, by "feeding the tube through itself" in such a way that we end up with a torus:

A torus, which is the simplest example of a compact Riemann surface, can also be constructed from a single square, by first stitching horizontal edges together…

...and then stitching the vertical edges (which have become circles) together:

The process of "curling up" a tessellation of the hyperbolic disc can also be thought of something like that. Corresponding edges of every copy of the tile get identified with one another, treated as if they were the same edge, the result being the same as taking a single tile and stitching its edges together in a particular way. The resulting object is a compact Riemann surface, but unlike a torus, one which can't be easily visualised. A torus is a "genus one" compact Riemann surface – the only kind which can't be created from tessellation of the hyperbolic disc[11]. All other compact Riemann surfaces, those with genus two or higher, can be created in this way. Roughly speaking, we can think of the genus of a surface as how many "holes" or "handles" it has, in the sense that a torus has one, and a novelty balloon shaped like a number 8 or a letter B would have two.

A compact Riemann surface in some ways resembles the surface of a sphere or a torus, in that a small enough inhabitant of such a surface would experience the small region in its immediate vicinity as being barely distinguishable from a similarly small piece of the plane. Likewise, the inhabitant can wander all over the surface, arriving back at its original location having never met an edge – a compact Riemann surface has no boundaries.

A compact Riemann surface with genus two or higher is, like a sphere, perfectly symmetrical – if you were a tiny inhabitant exploring such a surface, every direction, from every point, would look exactly the same. The key difference is that a sphere is an object of "constant positive curvature", whereas the hyperbolic disc and any of the compact Riemann surfaces it curls up into are objects of constant *negative* curvature. This is what makes them impossible to visualise or to construct in three-dimensional space. And, unlike the sphere, there are infinitely many compact Riemann surfaces. All spheres have exactly the same geometry (apart from size). These different compact Riemann surfaces are as geometrically distinct from each other as a sphere and a torus.

Although we can't visualise compact Riemann surfaces as objects in three-dimensional space, it is possible to produce images which are based on aspects of their geometry. So what you see here are not images of compact Riemann surfaces, but something more like "shadows" cast by slices of them, perhaps giving you a vague sense of what they might look like if you *could* see them:

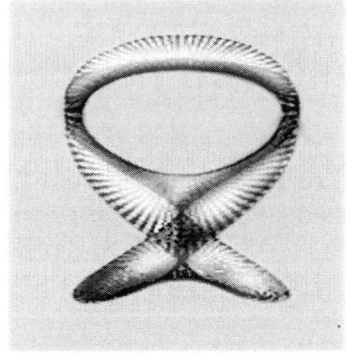 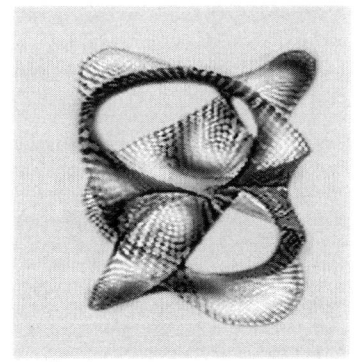 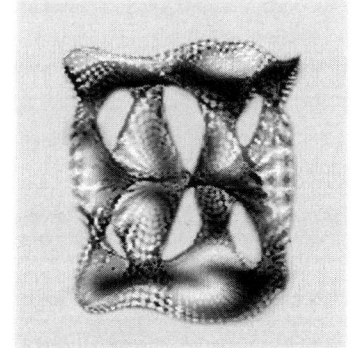

As was made clear back in Chapter 27, just because you can't put together a three-dimensional image or model of a mathematical object doesn't mean that you can't calculate a spectrum for it. Compact Riemann surfaces have associated with them a particular operator called the *Laplace–Beltrami operator*, jointly named after the late 18th century mathematician Pierre-Simon Laplace, and Eugenio Beltrami who adapted one of his mathematical innovations some decades later. Calculating the spectrum of eigenvalues of a Riemann surface's Laplace–Beltrami operator (there will be infinitely many eigenvalues), we end up with a representation of what is effectively "the sound you'd get if you could make one of these surfaces and hit it". As difficult as it might be to imagine, the physical theories of waves, vibrations, acoustics, *etc.* can all be extended from the familiar world of three-dimensional objects to this exotic world of abstract mathematical surfaces.

THE SPECTRAL GEOMETRY OF COMPACT RIEMANN SURFACES

Spectral geometry, you might recall from Chapter 27, is the study of how we can relate the geometry of an object to its spectrum. So a spectral geometer would naturally ask: can we relate the shape of a compact Riemann surface to its spectrum? It turns out that we can, using the Selberg Trace Formula:

The left-hand side of the STF (the part before the "=") concerns the spectrum of the surface. The r_i (involved in an infinite sum, as indicated by the "Σ") represent the various frequencies, or eigenvalues. Think of those as the individual tones, infinitely many of which make up the overall sound the surface would produce if you could "make one and hit it".

The right-hand side of the STF relates to the geometry of the surface. Imagine you are a tiny inhabitant on a perfectly smooth sphere with gravity but no friction. If you roll a ball in any direction, it will roll around the sphere in the biggest possible ("equator-sized") circle and come back to you. It doesn't matter where you roll it from or in which direction. These equator-sized circles are called the *geodesics* of the sphere and they all have the same length.

A sphere is among the simplest of surfaces. Others have more interesting sets of geodesics, usually not all of the same length. For example, imagine rolling a ball around on a torus (with a similar sort of gravity):

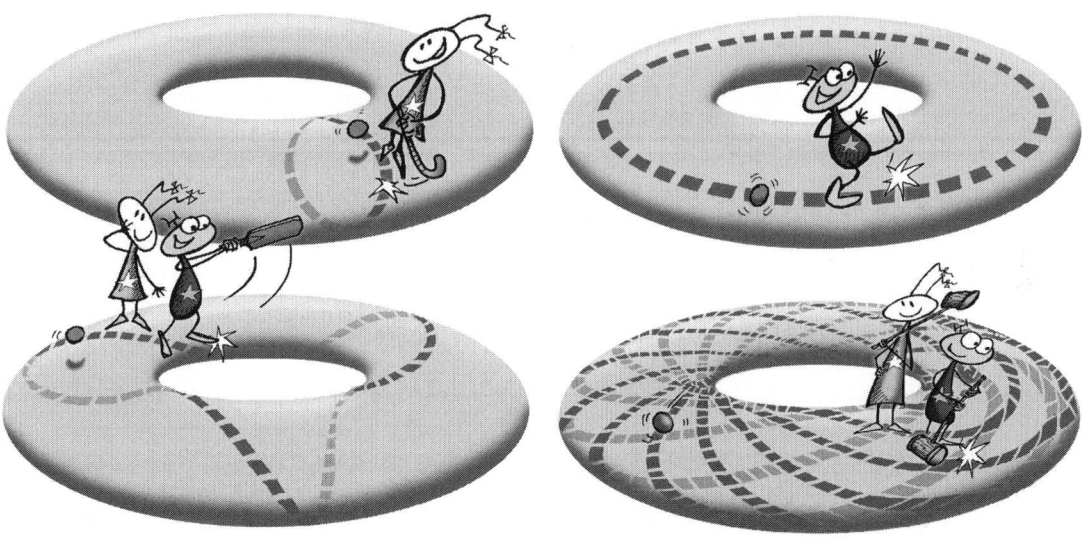

There are geodesics that involve going "through the hole", some that involve going "around the hole" and others that involve a combination of the two. The nature of the geodesic depends on which direction the ball is rolled. As shown in the last picture, some directions will produce *closed* geodesics – the ball will eventually return to its starting point, heading in exactly the same direction it started off in, hence repeating the same path endlessly. But others will produce geodesics which never repeat themselves, eventually covering the torus densely with their tracks:

Closed geodesics are also known as *periodic orbits* in this dynamical context, and their lengths are called *periods*, rather like the "period" in which a planet orbits the sun (although that's usually expressed as a duration rather than a distance). It's these periods which show up on the right-hand side of the STF, being directly related to the $\mathcal{N}\{P\}$'s.

So, the Selberg Trace Formula is a tool for use in spectral geometry. It expresses a relationship between the spectrum of eigenvalues of a compact Riemann surface and the lengths of that surface's periodic orbits. But what does this have to do with the spectral interpretation of the Riemann zeta zeros?

AN UNCANNY RESEMBLANCE

The answer is that the Riemann zeros show up in another formula, known as the *Riemann–Explicit Formula* [12], which (to mathematical eyes) looks suspiciously similar to the STF. And the role played by the zeros in this formula (they're denoted by the Greek letter γ) corresponds to the role played by the eigenvalues of the Riemann surface in the STF.

$$\sum_{\gamma} h(\gamma) = h\left(\frac{i}{2}\right) + h\left(-\frac{i}{2}\right) - g(0)\log\pi$$

$$+ \frac{1}{2\pi}\int_{-\infty}^{+\infty} h(r)\,\frac{\Gamma'}{\Gamma}\left(\frac{1}{4}+\frac{1}{2}ir\right)dr - 2\sum_{n=1}^{\infty}\frac{\Lambda(n)}{\sqrt{n}}\,g(\log n)$$

This formula may not look that much like the STF to you (see page 14), but once you understand the basics of both formulas, their striking similarity becomes evident.

It was first published in 1952 by André Weil, extending Riemann's explicit formula (the one we met back in Chapter 20). Riemann's formula, you might recall, related the Riemann zeta zeros on one side, and the prime numbers on the other side. Weil extended this in such a way that it would work not just for our familiar number system, but also for a whole range of other "number systems" (each with its own zeta function and generalised "prime objects"). We won't worry about that – it would require years of study to begin to understand these other number systems. The important thing for us is that when Selberg published his trace formula in 1956, it was immediately seen to have many similarities in form to Weil's explicit formula. No one has yet been able to directly relate them in a mathematically rigorous way – the resemblance is still "conjectural". So, based on appearances, it's widely believed that in some deep sense which we haven't quite grasped yet, these two formulas must have some common origin. The situation is something like the discovery of two paintings which stylistically appear to be the work of the same artist, but where this has yet to be shown convincingly to be the case.

Something else the two formulas have in common is the extent to which they're recognised as works of mathematical brilliance and of great beauty. Like the STF, the Riemann–Weil Explicit Formula (RWEF) is a kind of mathematical "Rosetta stone" which links previously unlinked mathematical structures in an initially surprising way.

This resemblance of the formulas means that we not only have the correspondence (based on their left-hand sides):

eigenvalues of a compact Riemann surface ⟷ Riemann zeta zeros

but also (based on their right-hand sides):

lengths of periodic geodesics of a compact Riemann surface ⟷ (logarithms of) prime numbers

Note that it's not the prime numbers themselves which correspond to the lengths of the periodic orbits, but *their logarithms*. At this stage you're probably not surprised by another appearance of logarithms in connection with prime numbers.

The resemblance of these two formulas was the first evidence to emerge suggesting that the zeta zeros are "spectral in nature"[13]. We've seen that there are many types of spectra, but in this particular context, the zeta zeros being spectral in nature suggests that they're "harmonics" – the vibrational frequencies of some (currently unknown) vibrating system. The STF-RWEF resemblance doesn't tell us what the unknown "vibrating system" is though, so despite there being cause to believe that the zeta zeros correspond to some kind of spectrum, we're no less in the dark as to what produces this spectrum. In other words, we're still left with du Sautoy's question "What's doing the vibrating?"

At first, the resemblance between the formulas might seem to suggest that there could be a very special compact Riemann surface for which the STF would exactly match the RWEF, but there are clear mathematical reasons why this cannot be[14]. In other words, although the heights of the Riemann zeta zeros *appear* to be the eigenvalues of something like a compact Riemann surface, we can be certain that they *aren't* the eigenvalues of any one particular compact Riemann surface.

Mathematicians don't generally deal with such vague imagery, but if you had to try to convey in simple terms what this resemblance is suggesting, it's this: it seems as if there exists some yet-to-be-discovered geometrical object, in some ways resembling a compact Riemann surface, which has a set of periodic orbits (or something like them) with lengths given by $\log 2$, $\log 3$, $\log 5$, $\log 7$, $\log 11$, $\log 13$, $\log 17$, *etc.*, and if you were to be able to somehow make one of these things and "hit it", then the resulting sound would be composed of infinitely many frequencies, these exactly matching the heights of the Riemann zeta zeros. To put it crudely, the Riemann zeros would correspond to tones (the sound of the thing), whereas the primes would correspond to loop lengths (the shape of the thing). No one has yet found such an object, although several researchers have been looking intently and have found things which *almost* fit the description.

So the resemblance of the two formulas provides "circumstantial evidence" for the heights of the zeta zeros being vibrational frequencies of something, and the "something" here appears to be a kind of abstract geometrical object whose geometry involves the sequence of prime numbers.

Before we go on to look at the second line of evidence for the spectral interpretation of the zeta zeros, we'll quickly consider the role of the STF in physics. This was established fairly soon after the formula's appearance in 1956, but it was initially seen as a separate matter from any connection between the STF and the Riemann zeta function. As we'll go on to see, though, this is no longer the case.

As part of the theory of compact Riemann surfaces, the STF is of interest to physicists studying something called *scattering*. This is a phenomenon which can involve either particles or waves, underpinned by a precise mathematical *scattering theory*. Early research relating the STF to scattering[15] can now be understood in relation to *chaos theory*, a branch of physics which began to emerge in the 1970s.

There's a lot of popular confusion about the nature of "chaos theory", so I shall attempt to clarify. Chaos theory is a sub-branch of the theory of dynamical systems. Had it been called something like "stochastic dynamical systems theory", it would probably have attracted a fraction of the attention that it has. Unfortunately, the somewhat dramatic, emotionally resonant word "chaos" seems to impair some people's ability to grasp the underlying concepts. It can create misleading expectations of what the theory is about (everyone imagining something slightly different) which wouldn't happen if a more neutral sounding name had been chosen.

I've also noticed that people sometimes ask me about "the chaos theory". The simple inclusion of the word "the" betrays a major misunderstanding. Compare:

the Big Bang theory		music theory
the theory of evolution	⟷	economic theory
the theory of relativity		analytic number theory

The left column contains "theories about the way things are" ("the" theories). The Big Bang theory claims (to put it very simply) that the universe began with an almighty explosion which came out of nowhere. The theory of evolution argues that biological life forms evolve to higher and higher levels of complexity as a result of random mutations and natural selection. The theory of relativity proposes that the universe is a curved four-dimensional "space-time continuum" and that all measurements must be made with respect to a "frame of reference" (but that there is no absolute frame of reference).

Although these three theories are widely accepted in 2013 and there's significant evidence to back them up, part of what makes them *theories* is that it's still possible, rationally, to doubt their accuracy. You can meaningfully believe or disbelieve in these theories.

The right column contains theories of a very different kind. You don't talk about "the music theory" as a theory of how the world works. You don't believe or disbelieve it. It's just a body of techniques and insights of use to musicians and composers. Similarly, "economic theory" is not "the economic theory" – it's just a body of techniques and ideas used by economists and investment bankers. (Within the world of economics, though, you'll find a number of competing theories of the first type – "the" theories, we could call them – which people can choose to believe or disbelieve). Analytic number theory is also a body of ideas (many proved) and techniques, not a speculation about the nature of things that people compile evidence in favour of or attempt to shout down.

Chaos theory is of the second type. There is no "the chaos theory" as some people have been mistakenly led to believe by badly written television documentaries and misguided hearsay. It has nothing to do with speculation that "the world is chaotic" or anything remotely like that. Instead, a community of dynamical systems theorists have been developing a body of mathematical models and techniques in order to understand and predict the behaviour of a particular class of dynamical systems which (for better or worse) have come to be known as *chaotic systems*. As the name suggests, these differ from the dynamical systems studied previously by the extent of their unpredictability, but it's still possible to develop a coherent mathematical theory that accounts for some aspects of their behaviour.

Here's a very simple example to illustrate the difference between chaotic and non-chaotic dynamical systems: Imagine a billiard table, with several balls set in motion. The paths they follow are entirely predictable, based on some simple physical laws involving "angles of incidence" and "angles of reflection". Now, imagine that we could freeze time at some point during this exercise and make an exact copy of the whole "system", placing it beside the original table. If we were then to unfreeze time, the two systems would spring back into motion, displaying identical behaviour in every detail.

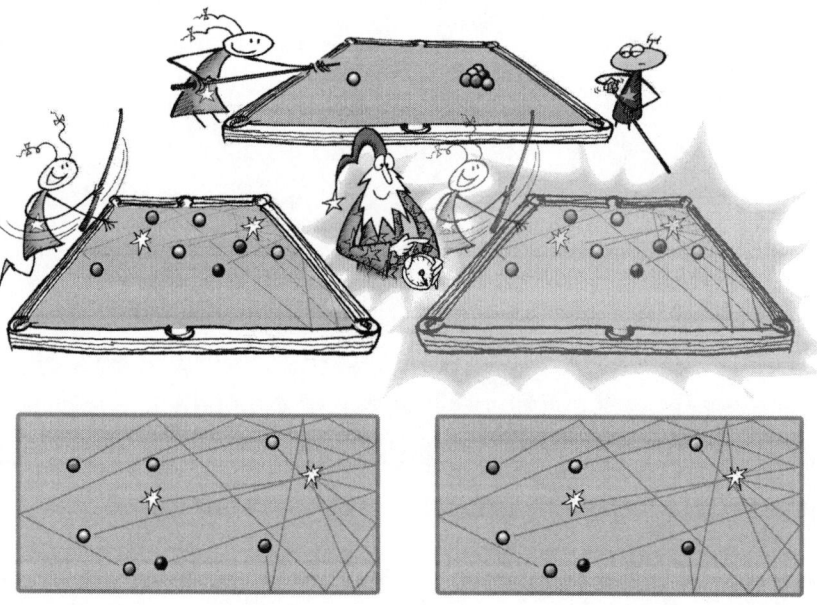

Now let's try that again, except this time we'll make some tiny change. We could move one of the balls in the "copied" system a fraction of a millimetre, say. Now unfreeze time. Because the copied system is so close to the original in every way, we'd expect its behaviour to closely match that of the original. And this is what happens. Over time, you'd see differences emerging, these being traceable back to the tiny displacement of the ball that we moved, but their emergence would be gradual. The word "gradual" is vague, of course, but it's possible to make a mathematically precise statement about this. The gradual nature of the emerging differences between the two versions would lead us to classify this as a non-chaotic system[16].

One simple way to create a chaotic system is to change the shape of the billiard table from a rectangle to one of these:

We carry out exactly the same procedure of freezing time, copying the system and moving one of the balls some tiny distance. In this case, we find that once time is unfrozen, the two systems rapidly start to show significant differences. Again, "rapidly" is a vague word, but it's possible to make this statement mathematically precise.

In mathematical language, chaotic and non-chaotic systems are distinguished by what gets called the "rate of divergence of nearby trajectories". In non-chaotic systems, the divergence between the systems is *linear*. In chaotic systems, this "divergence" is *exponential* (the difference between the two slightly different systems grows in a way related to the acceleration of the ladybird we met in Chapter 9 when the number $e = 2.718...$ was being introduced). To put it simply, a small change in a non-chaotic system leads to a small change in the way it evolves, while a small change in a chaotic system leads to huge change in the way it evolves. In this context, chaos theorists talk about "sensitivity to initial conditions".

We'll be coming back to it in Chapter 30, but that's really all we need to know about chaos theory at this point.

So if Weil's formula relating Riemann zeta zeros and prime numbers is truly (not just apparently) related to the Selberg Trace Formula, then we'd have a situation where the zeros could be linked to the physics of scattering and chaotic dynamics. This would be surprising, since the zeros were discovered as part of an investigation into the inner workings of the number system, an activity far removed from physics...or at least until relatively recently (as we'll soon go on to see).

Chapter 29

the spectral interpretation: Exhibit B

Now on to the second line of evidence for the spectral interpretation. This evidence originated in the 1970s and by the 1990s had become so extensive that any remaining doubts about the spectral interpretation of the Riemann zeta zeros had evaporated.

There's a relatively obscure branch of mathematics known as *random matrix theory* (I'll abbreviate this as "RMT"). Although it has become more widely known and applied in recent years, most mathematicians in 2013 are no more than vaguely aware of it. It was developed in the first half of the 20th century, motivated by certain problems in atomic physics.

As explained in Chapter 27, the simplest mathematical structure to produce a spectrum is a square matrix. These are regularly used in physics. Remember, square matrices have the same numbers of rows and columns:

$$\begin{bmatrix} 2 & -4 & 3 & 5 \\ -1 & -4 & -1 & 27 \\ 0 & 8 & 2 & 0 \\ 7 & 11 & 0 & -6 \end{bmatrix} \quad \begin{bmatrix} 1 & 0 & 0 \\ 0 & 1 & 0 \\ 0 & 0 & 1 \end{bmatrix} \quad \begin{bmatrix} 1.5 & 9.4 & -0.1 & 4.9 & 8.1 \\ 0.2 & 1.6 & 2.5 & -7.0 & 7.8 \\ -3.7 & -4.4 & -8.1 & -4.3 & 3.5 \\ 6.6 & 6.8 & -7.0 & -0.8 & 8.7 \\ -0.8 & 1.0 & 11.2 & 2.2 & -9.9 \end{bmatrix} \quad \begin{bmatrix} 3-i & 4i \\ 6 & 7+3.5i \end{bmatrix}$$

You might imagine that random matrix theory would deal with "random matrices". But there's actually no such thing as a "random matrix". It's more subtle than that.

In RMT you work with the set of all possible square matrices of some particular type and some (usually "large") specified size. There are a number of classifications of square matrices (but we won't worry too much about them): *symmetric* matrices, *skew-symmetric* matrices, *Hermitian* matrices, *unitary* matrices, *orthogonal* matrices, *symplectic* matrices, *positive-definite* matrices, *etc*. These classifications are all defined in terms of precise properties concerning the way the various "entries" of a given matrix interrelate. So you can talk about the collection of all possible "unitary 40×40 matrices" or the collection of all possible "symmetric 250×250 matrices with positive real number entries", for example. You can then imagine choosing one of these matrices at random from the collection... except, again, it's a bit more subtle than that.

Some of the matrices in our collection could be more heavily *weighted* than others (this weighting being based on the matrix entries). The idea is that the more heavily weighted a matrix is, the more likely it is that our process of "random selection" will produce that matrix. One way of thinking about this is to imagine all of the possible matrices in the collection as people who've bought tickets in a lottery, and where almost everyone has bought more than one ticket. The more heavily weighted matrices correspond to the people who've bought more tickets (the weighting is directly related to the number of tickets held) so they're the matrices which are most likely to "win" in a random draw.

The process responsible for "weighting" the matrices in the collection is called a *probability distribution*. (We met a well known example of a probability distribution in Chapter 22 – the Gaussian distribution – but that concerned the weighting of real numbers rather than matrices.) Some choices of matrix are more probable (that is, have a higher probability of being chosen) than others and the process of weighting is how these probabilities are quantified.

We have to imagine carrying out the random lottery draw over and over again and watching carefully which matrices come up. Of course *any of them* can come up (they've all got *some* weight, that is "at least one ticket") but some will come up more often than others. A collection of matrices of a certain type together with an appropriate probability distribution for weighting them, is known as a *random matrix ensemble*. The one we'll be concerned with is known as the *Gaussian unitary ensemble* (GUE) of matrices of a given size. You can talk about the GUE of 10×10 matrices, of 370×370 matrices or of any size you like. The details needn't concern us, we just need to bear in mind that (1) there's a certain type of square matrix called a Hermitian matrix, (2) for any given matrix size, we can discuss the set of all possible Hermitian matrices of that size, and (3) the GUE involves a very particular scheme (probability distribution) for supplying a "weight" to each of these Hermitian matrices. You might be wondering how an individual weight could be associated with every single matrix in an infinite collection – wouldn't that require an infinite amount of time? The answer is no, because the probability distributions which are studied on these sets of matrices are always defined in terms of functions on the matrix entries – that is, they're given in terms of a formula, so given a matrix in the collection, its weight can be calculated directly from its entries [1].

We know that each of these matrices will have a spectrum of eigenvalues, here shown graphically for an example of an 11×11 Hermitian matrix (such matrices, despite their complex entries, are known to have all real eigenvalues):

9.1	9.2+9.1i	-6.8+3.1i	9.4-9.2i	9.1+6.9i	-0.2+8.6i	6.0+3.5i	-7.1+5.1i	-1.5+4.8i	8.3-2.1i	5.8+3.1i
9.2-9.1i	-0.9	-8.3-7.8i	-5.4+9.2i	8.2-9.9i	-6.9+5.4i	6.5+6.3i	0.7+7.3i	9.9-8.3i	-8.4-2.0i	-1.1-4.8i
-6.8-3.1i	-8.3+7.8i	5.5	8.6-3.7i	-7.4+0.5i	1.3-6.6i	-0.6+2.0i	-9.7-4.7i	-3.2+3.0i	-6.7+3.7i	5.8+4.9i
9.4+9.2i	-5.4-9.2i	8.6+3.7i	2.3	-0.5+5.1i	-2.9+5.0i	6.6-2.3i	1.7+1.3i	0.9-8.4i	8.3-8.9i	-4.2+0.6i
9.1-6.9i	8.2+9.9i	-7.4-0.5i	-0.5-5.1i	9.1	0.9+6.2i	-7.2-5.1i	-7.0+8.5i	-4.8-3.0i	6.8-6.0i	-4.9-4.9i
-0.2-8.6i	-6.9-5.4i	1.3+6.6i	-2.9-5.0i	0.9-6.2i	-7.6	-0.0+5.0i	9.1-4.8i	-3.1+0.1i	1.7+3.9i	-5.5-7.8i
6.0-3.5i	6.5-6.3i	-0.6-2.0i	6.6+2.3i	-7.2+5.1i	-0.0-5.0i	-0.2	-1.0-4.4i	2.9+3.5i	4.1+3.1i	5.0-6.7i
-7.1-5.1i	0.7-7.3i	-9.7+4.7i	1.7-1.3i	-7.0-8.5i	9.1+4.8i	-1.0+4.4i	9.0	-9.3+5.3i	-1.2+5.9i	-2.3-6.2i
-1.5-4.8i	9.9+8.3i	-3.2-3.0i	0.9+8.4i	-4.8+3.0i	-3.1-0.1i	2.9-3.5i	-9.3-5.3i	-9.0	-8.0+3.8i	6.4-3.6i
8.3+2.1i	-8.4+2.0i	-6.7-3.7i	8.3+8.9i	6.8+6.0i	1.7-3.9i	4.1-3.1i	-1.2-5.9i	-8.0-3.8i	4.1	-9.3-4.4i
5.8-3.1i	-1.1+4.8i	5.8-4.9i	-4.2-0.6i	-4.9+4.9i	-5.5-7.8i	5.0+6.7i	-2.3+6.2i	6.4+3.6i	-9.3+4.4i	-6.5

So we repeatedly run a random draw on our matrix ensemble and keep track of all the spectra of the "winning" matrices. Suppose we collect a large number of these and then run a statistical analysis on them (as a market researcher might on a large number of consumer purchases, or a biologist might on a large number of eels).

What trends can we detect in the spectra that tend to show up? Trends can concern (among other things) *distances between pairs of eigenvalues* – that could be just between neighbouring pairs, or between *all* pairs. We can consider average, maximum or minimum distances. There are many such trends which can be observed, checked for, speculated about or even mathematically proved. This is the sort of thing that random matrix theory deals with – the various statistical signatures of different types of matrix ensembles.

RANDOM MATRIX THEORY IN PHYSICS

In the mid-1970s, one of the world's leading authorities on random matrix theory was the physicist Freeman Dyson, at that time based at the Institute for Advanced Studies (Princeton, New Jersey). He'd been successfully exploring aspects of the subatomic world using RMT since the early 1960s, following on from the work of Eugene Wigner who'd first developed the approach in the mid-50s. This involved attempting to learn about certain types of atomic systems via their spectra by considering classes of appropriately large random matrices whose spectra display similar trends. As part of this, Dyson was interested in looking at the distances between eigenvalues in the spectra produced by various types of random matrix ensembles. The hope was that the overall statistical behaviour of these distances might involve certain identifying spectral signatures (think of fingerprints) which could then be linked to specific physical phenomena. Subatomic systems with common features could then be grouped together according to the types of matrices which tend to match their "spectral behaviour".

Wigner had originally suggested that random matrix ensembles might be useful in subatomic physics in 1955[2]. Matrices had been heavily involved in quantum mechanics for some decades by then. Although physicists had used them previously[3], the significant role of matrix mathematics in quantum mechanics (along with the unavoidable involvement of complex numbers) is largely what set it apart from previous physics.

When introducing matrices in Chapter 27, I mentioned that you can multiply pairs of them (if their sizes are "compatible"). Two square matrices of the same size can *always* be multiplied. However, if you multiply them "the other way round", you usually don't get the same answer:

$$\begin{bmatrix} 2 & -1 \\ 3 & 7 \end{bmatrix} \begin{bmatrix} -4 & 0 \\ 2 & 5 \end{bmatrix} = \begin{bmatrix} -10 & -5 \\ 2 & 35 \end{bmatrix} \qquad \begin{bmatrix} -4 & 0 \\ 2 & 5 \end{bmatrix} \begin{bmatrix} 2 & -1 \\ 3 & 7 \end{bmatrix} = \begin{bmatrix} -8 & 4 \\ 19 & 33 \end{bmatrix}$$

$$\begin{bmatrix} 1 & 0 & 0 \\ 0 & 1 & 1 \\ 0 & 0 & 0 \end{bmatrix} \begin{bmatrix} 1 & 0 & 0 \\ 0 & 0 & 0 \\ 0 & 0 & 1 \end{bmatrix} = \begin{bmatrix} 1 & 0 & 0 \\ 0 & 0 & 1 \\ 0 & 0 & 0 \end{bmatrix} \qquad \begin{bmatrix} 1 & 0 & 0 \\ 0 & 0 & 0 \\ 0 & 0 & 1 \end{bmatrix} \begin{bmatrix} 1 & 0 & 0 \\ 0 & 1 & 1 \\ 0 & 0 & 0 \end{bmatrix} = \begin{bmatrix} 1 & 0 & 0 \\ 0 & 0 & 0 \\ 0 & 0 & 0 \end{bmatrix}$$

As with the multiplication of variables in elementary algebra (expressed like "x = yz"), multiplication signs tend not to be used between pairs of multiplied matrices.

To describe this situation, mathematicians say that the multiplication of matrices is *noncommutative*. When you multiply real or complex numbers, the order of multiplication doesn't matter, you get the same result either way:

$2 \times 5 = 10$	$5 \times 2 = 10$
$7.23 \times 0.62 = 4.4826$	$0.62 \times 7.23 = 4.4826$
$(4 - i) \times (3 + 2i) = (14 + 5i)$	$(3 + 2i) \times (4 - i) = (14 + 5i)$

Multiplication in $\mathbb{R}$ and $\mathbb{C}$ is *commutative*. Physics which can be expressed in terms of $\mathbb{R}$ and/or $\mathbb{C}$ is able to go quite a long way in describing the workings of physical reality, but once you descend into the "quantum realm" (where you're dealing with size scales associated with subatomic particles) these descriptions will no longer suffice. To get some sense of the scale involved here, imagine taking 1 centimetre, then a tenth of that (1 millimetre), then a tenth again, and again five or six more times – extremely small by human standards, but still a positive size. At that scale, the physical world seems to be *intrinsically noncommutative* and the only accurate mathematical descriptions of it involve noncommutativity. This is why matrices came to be adopted as an important tool by quantum physicists.

Not surprisingly, subatomic physics gets more complicated as you add more particles to the system you're considering – there's an increasing number of interactions to keep track of. The very simplest atoms can be almost completely described in terms of mathematical laws, but heavier ones (those with more particles) are simply not manageable in this way. Applying mathematical techniques to calculate the spectra of small atoms sometimes works, but as more particles get involved this becomes out of the question. Wigner's innovation was to suggest that you could create mathematical models for studying the spectra of "heavy nuclei" by using appropriate random matrix ensembles. Remember, these involve the random selection of matrices – no individual matrix can be said to be a "random matrix", just as no individual number (without a clear context for doing so) can be meaningfully called a "random number". The language is potentially confusing, but notice that we talk about "random matrix ensembles" rather than "ensembles of random matrices".

A very much simplified account of Wigner's reasoning would be as follows: The simplest nuclei have spectra with some sort of regularity to them (there are reasonably simple laws governing the distributions of their eigenvalues). But once the nuclei involve enough particles, all regularity disappears and the spectra begin to look *as if* they were random (they're the result of fixed, deterministic physical laws, but they *look* random). The idea, then, is to treat them as if they *were*, in some sense, the spectra of randomly chosen matrices, and see what can be deduced.

There are different "styles" of randomness, though, depending on the probability distribution you're using to define your random matrix ensemble (this is my own choice of word – "styles" isn't a formal mathematical term). Remember, our choice of probability distribution provides the system for weighting the various matrices in the ensemble, establishing how much more likely some matrices are to show up than others. The details of this system are what give the resulting randomness its particular "style". The style is then evident in the trends you see when you look at the spectra of many different matrices chosen from the ensemble. So GUE spectra tend to have certain "statistical fingerprints" which differ from those of, say, GOE (Gaussian orthogonal ensemble) spectra or CSE (circular symplectic ensemble) spectra.

We examine our complicated "nuclear spectrum", then, and attempt to deduce which "style" of randomness it displays (based on certain properties involving the spacings between eigenvalues). In other words, we're considering *which ensemble it looks like it comes from*. This comes down to statistically analysing eigenvalues.

It turns out that you can indeed arrive at very useful classifications of, and deductions about, nuclear spectra with this approach. Wigner's idea does work, and very well. As part of this programme of research, Dyson produced a formula describing the statistical distribution of gaps between eigenvalues in spectra from the random matrix ensemble which most concerns us, the Gaussian unitary ensemble. For any eigenvalue in such a spectrum, the formula tells you the likelihood of finding another eigenvalue within any given distance from it.

Among other things, Dyson's formula shows that the eigenvalues have a tendency to "repel each other". Gaps between nearest-neighbour eigenvalues tend to be larger than you'd expect if the eigenvalues were just randomly distributed across the spectrum, entirely independently of each other.

This all may be very interesting, but what does it have to do with the Riemann zeta zeros?

A FORTUITOUS MEETING

The connection came about very suddenly in 1972 when a young number theorist called Hugh Montgomery was visiting the IAS in Princeton and was informally introduced to Dyson over tea. When Dyson asked what he was working on, Montgomery explained that he'd been looking at the spacings between the nontrivial zeros of the Riemann zeta function. In fact, he explained, he'd come up with (but not proved) a formula which described the statistical distribution of these gaps. When Montgomery described the formula, Dyson suddenly became very excited and surprised Montgomery (who'd not even heard of random matrix theory, being predominantly interested in number theory) by exclaiming *"That's the form factor for the pair correlation of eigenvalues of random Hermitian matrices!"* [4].

Hugh Montgomery and Freeman Dyson never again met or worked together, but this fortuitous meeting suddenly revealed to the mathematical community that the nontrivial zeros of the Riemann zeta function appear to have a "statistical fingerprint" shared by spectra of large GUE matrices. In this way, a deep connection was forged between analytic number theory and random matrix theory. Although Montgomery's "pair correlation conjecture" hadn't been proved (in fact, it still hasn't[5]), there was sufficient numerical data available on the Riemann zeta zeros at that time to back it up convincingly. This connection with RMT compounded the suspicion which had been fuelled some years earlier by the "Selberg–Weil resemblance" – that the zeta zeros are, in some sense, *the spectrum of something*.

Hugh Montgomery went on to conjecture that the Riemann zeta zeros and GUE spectra share not just this one particular statistical property, but *all possible* statistical properties – that they are *statistically identical*. In the 1980s, Andrew Odlyzko, who (as of 2013) has calculated hundreds of millions of Riemann zeta zeros[6], performed some epic computations which backed up Montgomery's so-called *GUE Hypothesis* (consequently, it's now sometimes called the *Montgomery–Odlyzko Law*). These computations eliminated even the tiniest hint of doubt that the Riemann zeros are intimately related to the spectra of GUE matrices. But despite the enormous body of empirical evidence and near certainty among mathematicians involved, this "law" remains an unproven conjecture (rather like the "laws" of physics).

The "RMT connection" tells us, very roughly, that the heights of the nontrivial zeros of Riemann's zeta function look like the eigenvalues making up the spectrum of a particular kind of "infinitely large random matrix". More precisely, it allows us to look at a certain stretch of the critical strip and say that *in that region*, the heights of the zeros look like they originate with a random matrix of a particular size (based on how far up the strip you're looking). For example, it's been pointed out[7] that up in the vicinity of the $100\,000\,000\,000\,000\,000\,000$th zero (the height of which up the critical line is approximately $15\,202\,000\,000\,000\,000\,000$), the nontrivial Riemann zeta zeros share the statistical fingerprints of eigenvalues from spectra of 42×42 matrices drawn from the (42×42) Gaussian unitary ensemble.

The zeta zeros can't meaningfully be described as "random" – they're fixed and computable. So rather like the primes, the zeta zeros *appear* random in some quite specific ways while at the same time being "carved in stone" and fundamental to the structure of reality.

Quite what all of this means is far from clear, but along with the Selberg–Weil resemblance, Odlyzko's extensive evidence for Montgomery's GUE Hypothesis has led to a near certainty that the nontrivial zeros of the Riemann zeta function are "spectral in nature", that is, that they correspond to a spectrum (of unknown origin).

chapter 30

quantum chaos

In Chapter 28, we saw evidence for the spectral interpretation of the Riemann zeta zeros in the form of an unexplained resemblance between the Selberg Trace Formula and the Riemann–Weil Explicit Formula. The STF, remember, concerns "compact Riemann surfaces", relating the geometry of each of these to its spectrum of eigenvalues (harmonics). The role of the zeta zeros in the RWEF corresponds to the role of the eigenvalues in the STF. Similarly, the role of the primes in the RWEF corresponds to the role of the "periodic orbits" or "closed geodesics" in the STF (see page 18).

I mentioned that Selberg's formula is of interest to physicists working in chaos theory. I also explained that the geodesic motion of a point particle on a sphere is very simple, always orbiting an equator-sized circle (an example of non-chaotic motion). At the other extreme, the geodesic motion of a point particle on the hyperbolic disc, while also appearing simple, is highly chaotic.

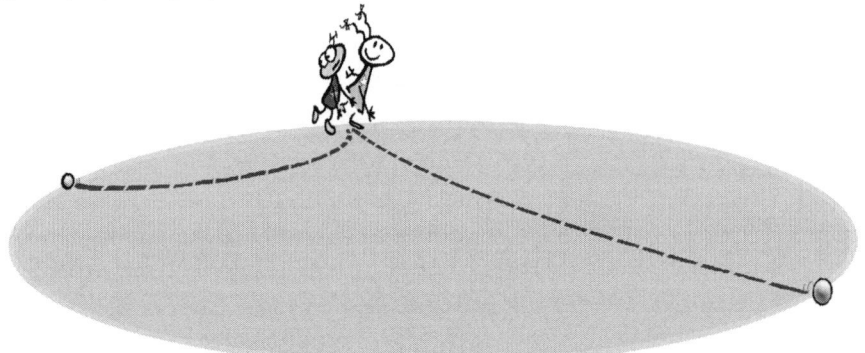

Recall from Chapter 28 that chaotic motion isn't about crazy, unpredictable zig-zaggings, it's about *sensitivity to initial conditions* and *rapidly diverging trajectories*. Here a tiny difference in the initial direction of motion leads to such a rapid divergence. The trajectories shown on the hyperbolic disc here are geodesics (the equivalent of straight lines in the plane).

Because of the way compact Riemann surfaces were described ("curlings up" of the hyperbolic plane), it can easily be deduced that the geodesic motion of point particles on those surfaces will share this chaotic property.

When investigating a particular dynamical system, chaos theorists will often study its periodic orbits. They might seek to relate these to a spectrum of eigenvalues associated with the system. In the case of point particle motion on a compact Riemann surface, this is exactly what the STF is concerned with, although it appeared some years before chaos theory came into being. (The word "chaos" first appeared in print with this specific meaning in 1975[1], although the theory has its roots in earlier research.)

Mathematical physicist Martin Gutzwiller published another "trace formula" in 1971[2]. The Selberg Trace Formula belongs primarily to mathematics, although applications in physics have emerged since its discovery. The Gutzwiller Trace Formula, on the other hand (although now of interest to some pure mathematicians) belongs primarily to physics. It's a useful tool for chaos theorists, relating the spectrum associated with a dynamical system to the lengths of that system's periodic orbits.

Selberg's formula, it turns out, can be thought of as just *one special case of Gutzwiller's formula*, where the dynamical system under consideration is simply a single point particle moving freely on a compact Riemann surface[3].

The next year, 1972, Hugh Montgomery met Freeman Dyson and the other current of evidence for the spectral interpretation was brought into awareness – the GUE Hypothesis explained in the previous chapter. Recall that Dyson was a physicist and that the GUE is part of random matrix theory which, although a branch of mathematics, historically originated with research in atomic physics.

The evidence was strengthened when Andrew Odlyzko, having heard Montgomery lecture on these matters in 1978[4], began his extensive programme of computation to calculate nontrivial Riemann zeta zeros at unprecedented heights. This vastly expanded body of data then allowed much more detailed comparisons to be made between zeta zero spacing and GUE eigenvalue spacing statistics.

The evidence Odlyzko produced, although slightly problematic in one detail which we'll come back to at the end of the chapter, was widely perceived by those involved in these matters as truly remarkable. Its implication was that the subtle spacing statistics of the zeta zeros so closely resemble those of the spectra of eigenvalues belonging to GUE matrices[5] that there can be no doubt that the zeta zeros are "spectral in nature".

So we have overwhelming evidence that the Riemann zeros are behaving like the spectrum of something, but still no idea as to what that "something" might be. The fact that the GUE was originally put to successful use by nuclear physicists hints at the possibility that there might be a connection between the zeta zeros and nuclear physics. Could the unknown "vibrating something" underlying the number system somehow involve this kind of physics, and if so, *what would that mean?*

It seems a strange possibility indeed that ideas from physics might be required to account for fundamental phenomena in the purest branch of pure mathematics. If they are, though, the trace-formula-related evidence suggests that it'll be *chaotic* physics which will provide the necessary clues, while the random matrix connection hints that it could involve subatomic (that is, *quantum*) physics. Now, despite a widespread misconception among non-scientists who lump them together as exotic new branches of physics which must therefore be somehow related, there's very little common ground between quantum mechanics and chaos theory. Chaos theory directly applies only to *classical* physics (that's how physics which predates quantum mechanics is now classified[6]). As physicist Michael Berry has explained:

> "*In classical mechanics, objects can move along infinitely many trajectories... This makes it easy to set up complicated dynamics in which an object will never retrace its path – the sort of behaviour that leads to chaos. But in quantum mechanics...the fine detail* [is blurred out], *smoothing away the chaos.*"[7]

However, with the emergence of the Gutzwiller Trace Formula came a new branch of physics which acts as a point of contact between quantum mechanics and chaos theory. To avoid the misleading term "quantum chaos", Berry suggested calling it *quantum chaology* (both names have since gone into circulation). It is here that the two currents of evidence for the spectral interpretation of the Riemann zeta zeros fully converge. In some very strange and unexpected way, it now seems certain that *the zeta zeros are related to quantum chaology.*

To "quantum chaologists", the GUE and its spectra are associated with a particular class of chaotic dynamical systems. Although classical, these systems have quantum mechanical counterparts which produce GUE-like spectra. This has led to the realisation that the Riemann zeta zeros collectively display a curious resemblance to the spectrum of a particular type of chaos-related quantum mechanical system. Notice that I've written "chaos-related" rather than "chaotic". This, as I've explained, is because a quantum mechanical system cannot display the properties of chaos, chaos only occurring in *classical* physics. We'll need to explore this quantum-classical distinction further before coming back to the zeta zeros and how this relates to them.

QUANTUM VERSUS CLASSICAL

Whereas classical mechanics typically studies the interactions of idealised "point-particles" in motion (think of billiards colliding), quantum mechanics is more like the study of the complex interactions of ripples on a pond. Although the latter study would still be considered part of classical physics, it's a type of "wave mechanics". Quantum mechanics is also a type of wave mechanics, although the "waves" in question are very far removed from the familiar ones which propagate across the surfaces of bodies of water.

Whereas the word "quantum" is perhaps most familiar as an adjective (usually preceding "physics" or "mechanics"), its original use was as a noun – a quantum of energy, for example, is the smallest quantity of energy that can be meaningfully considered in a particular physical context. It's often described as a "packet" or "bundle" (I'd suggest "blob"). Physics experiments at the end of the 19th century first revealed the unexpected fact that certain physical quantities like *charge* and *action*[8] are "granulated" into quanta (that's the plural).

The easiest way to picture this is in terms of light, the most familiar and accessible form of electromagnetic energy. Imagine a laser producing a single frequency of, say, green light, the intensity of which is controlled by a knob something like the "dimmer switches" now fairly common in domestic lighting. Turning the knob clockwise increases the intensity of the light, while turning it anticlockwise reduces it. Everyday experience suggests that we can turn the light intensity down towards zero as close as the precision of the equipment will allow. However small, we should (theoretically) be able to halve it, and again, and again... But this isn't how things work. For any given frequency, *there is a smallest amount of light energy possible*, so it's as if our knob doesn't turn continuously, but in a succession of very tiny "steps". Nothing is possible between zero (no energy whatsoever) and a single quantum of energy.

This counterintuitive discovery is related to what's known as the *wave-particle duality*. Experiments in classical optics (involving lenses and mirrors, reflection and refraction) all suggested that light behaves like a wave, whereas this "quantised" property of light meant that other experiments could be devised which seemed to indicate that light behaves instead like a stream of particles or *photons* (each possessing a single quantum of energy). The physics community has come to accept that light somehow manages to be both, depending on how you look at it.

A helpful analogy (if not taken too literally) involves a dripping water tap. Suppose we turn a tap to reduce a gush of water down to a narrow, continuous stream. It might seem that by gently turning the tap ever closer to its "closed" position, we could narrow the stream of water to an ever thinner thread of flowing liquid. But we know that this isn't what happens – instead, there comes a point where the water is reduced to a sequence of separate drops, all (under ideal conditions) of the same size [9]. We could think of quanta as being something like these drops. The volume of water released from an ideal dripping tap will always be a whole number multiple of the volume of a single drop, just as the quantity of light energy emitted from our single-frequency laser source (at any intensity) will always be a whole number multiple of a single quantum of energy.

PLANCK'S CONSTANT

Max Planck's 1889 discovery of the quantised nature of *action* (that's energy × time) led him to define what's now known as *Planck's constant*, written h, a single quantum of action (in many situations, the closely related constant $\hbar = h/2\pi$ is used). As part of his discovery, Planck realised that the energy of a photon is equal to h multiplied by the frequency of the light (or other type of electromagnetic energy) involved.

The discovery of this constant led Planck to formulate a set of "natural units" for five fundamental quantities: mass, time, length, charge and temperature (called *Planck mass, Planck time, etc.*). The *Planck length* is probably the best known of the "Planck units". It's what you get if you start with 1.6 centimetres, then go down to 1.6 millimetres, then continue to take tenths another 32 times. That'd be written "1.6×10^{-35} m" in scientific notation. It's extremely small by human standards, in fact it's several orders of magnitude smaller than the smallest known subatomic particles. The

Planck mass is about 22 millionths of a gram – again, tiny relative to our everyday experience, but this quantity of matter would be something a chemist could happily work with. The *Planck time* is the amount of time it takes light (which travels at about 300000 metres per second) to travel a single Planck length, so we're again dealing with an unimaginably tiny quantity (in this case of duration).

By definition, then, *light travels at 1 Planck length per Planck time*. In other words, if we adopt Planck units, then the speed of light, one of the fundamental physical constants, becomes equal to 1. The beauty of Planck units, and why they can be called "natural", is that if we measure all quantities with them, then the values of five fundamental constants (the "reduced Planck constant" $\hbar$, the speed of light c, *Boltzmann's constant* k_B, the *gravitational constant G* and the *Coulomb constant* k_e)[10] all become equal to 1 and thus vanish from many key equations. For example, the famous $E = mc^2$ becomes simply $E = m$ (the equivalence of energy and mass, measured in units of Planck energy and Planck mass [11]). Importantly, unlike metric, imperial and other systems of units, the Planck units are not arbitrarily chosen or culturally rooted.

The five primary Planck units, as well as numerous other "derived" Planck units (Planck energy, area, volume, density, power, *etc.*) can all be defined in terms of the five fundamental constants listed above. For example, the Planck length is the square root of $\hbar G/c^3$. Importantly, all five primary Planck units are defined such that if we were able to make h (and hence $\hbar$) shrink towards 0, they would also shrink towards 0. Recall that the amount of energy in a quantum of light (a photon) is h times the light frequency, so letting h shrink towards 0 would also shrink the size of our quanta ("blobs") of electromagnetic energy.

The Planck length, incidentally, can be understood as *the smallest length which can be measured*. As Roger Penrose explains [12], it's the length scale "*at which the so-called 'quantum fluctuations' in the very metric of space-time should be so large that the normal idea of a smooth space-time continuum ceases to apply.*" It's not just that we aren't sufficiently advanced or that our instruments aren't sophisticated enough to measure beyond this resolution – the laws of physics rule it out altogether, rendering the possibility meaningless. The universe looks very different at the Planck scale.

Planck's constant is central to the workings of quantum mechanics. Consequently, many quantum mechanical formulas contain the symbol h (or, more commonly, the reduced version $\hbar = h/2\pi$). Probably the most famous of these is the *Schrödinger equation*[13]. Associated with a quantum mechanical system there will be a spectrum of energy levels (this was discussed in Chapter 27) and for some relatively simple systems, the Schrödinger equation can be used to compute exactly what that spectrum will be.

$\hbar$ also shows up in the *Heisenberg uncertainty principle*[14], a mathematical principle which sets limits on how precisely we can measure certain quantities relative to others. These come in "trade-off" pairs, such as position-momentum and time-energy. So, for example, the more precisely we can measure the position of a particle, the less precisely we can measure its momentum, and *vice versa*. But the *extent* of the uncertainty which this involves is controlled by the size of h. Smaller h (or $\hbar$) means less uncertainty, that is, greater precision possible in both measurements simultaneously.

I mentioned in Chapter 29 that the mathematics used in quantum physics is largely noncommutative. There's even a thing called a *commutator* which measures "how noncommutative" certain things are. If x_i is a "position operator" and p_j is a "momentum operator" (these things are in some ways like infinite matrices), then the commutator $[x_i, p_j]$ is defined as $x_i p_j - p_j x_i$. This would be 0 if these operators commuted…but they don't. A basic quantum mechanical result tells us that $[x_i, p_j] = i\hbar$ (where i is the "imaginary unit" such that $i^2 = -1$). So the size of h controls "how noncommutative" the pairs of operators are. For smaller h (and hence smaller $i\hbar$) the operators would start to approach something like commutativity, which is the classical situation.

Crudely speaking we can think of the fundamental new idea in quantum physics as being the fact that physical reality is "granulated" in a certain sense, rather than continuous (as classical physics had assumed). And in a similar spirit, we can think of Planck's constant h as controlling "the size of the granules".

These ideas surrounding Planck's constant are fascinating. We're looking at a sort of boundary between two different worlds, the classical and the quantum. Although these are obviously both parts of a single whole, we don't yet have a clear picture of how it all connects up. But physicists have been working on this – there's now something called *semiclassical physics* which, in some sense, deals with the border or interface between the two worlds.

SEMICLASSICAL PHYSICS

Because of the central role of Planck's constant in quantum mechanical calculations, it's inevitable that physicists would eventually try to imagine a *different* quantum physics where the constant happens to be some other value. The "true" Planck constant was found with a combination of theoretical and experimental work (and since repeatedly confirmed), but it seems arbitrary – why *this particular* value? Of course you can't actually *change* the Planck constant (just as you can't change the strength of gravity) but you *can* mathematically describe and work with the physics you'd end up with if you could. Quantum mechanical theory will continue to "work" with other positive values of h – it will just fail to accurately describe this universe.

And if we imagine quantum physics for smaller and smaller Planck constants, gradually letting this quantity shrink down towards zero in all relevant equations and calculations, then we end up back in the world of classical mechanics. In particular, the Planck length, the length below which it is impossible to measure, will shrink towards zero, meaning that *there's no limit to how small measurements of length can meaningfully be* (similar situations applying for other quantities). This is a crucial assumption which classical mechanics had taken for granted prior to the quantum revolution.

Returning briefly to our dripping tap analogy, taking h down towards 0 would be the equivalent of making the sizes of the drops shrink down towards zero. In an idealised sense, you should be able to see that when the drop sizes approach zero, we're effectively returning to our "pre-quantised" (or classical) continuous stream of water.

Whereas quantum mechanics is a type of wave mechanics, classical mechanics is said to be "ray-based". Things like billiard balls travel in straight lines ("rays") and bounce off each other making clean, sharp angles. Quantum mechanics is "blurrier", involving interacting waveforms (in dimensions impossible to visualise). But if you take the all-important Planck constant down towards zero, the waves effectively become rays, and everything "goes classical" – all the counterintuitive weirdness of the quantum world evaporates and we're back in the comfortingly familiar world of classical physics.

Semiclassical physics involves studying what happens when you imagine the Planck constant shrinking down towards (but never quite reaching) zero. You're approaching the transition from the quantum back to the classical realm, but never quite reaching it. This is what's known as the *semiclassical limit*.

In this way, we can transform (the mathematical model of) a quantum dynamical system into (the mathematical model of) a classical one. We start shrinking the Planck constant, so the system's behaviour starts to shift (this constant being one of the components in the underlying mathematics). By letting h dwindle towards zero in all of the relevant formulas, our quantum dynamical system mathematically transforms into a classical dynamical system. This is said to be the classical system "underlying" our quantum system.

There's another process which goes in the other direction – a process which, given a classical system, will produce a quantum system which *it* underlies. This is called *quantisation*, but fortunately we won't need to worry about it:

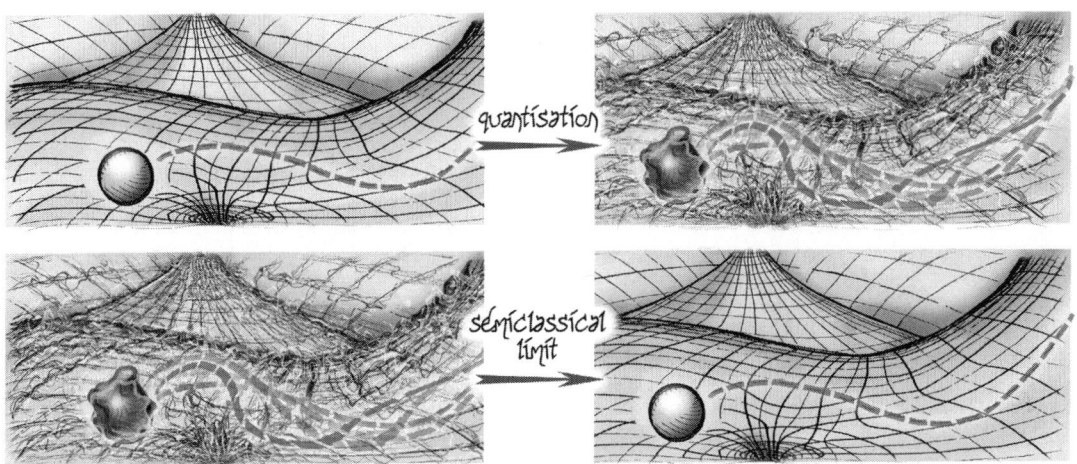

The quantisation of a classical system (upper-left) is a quantum system (upper-right). Applying the semiclassical limit to this takes us back to the original classical system (lower-right). Unlike the simplicity of the semiclassical limit ("$h \to 0$"), quantisation is much more problematic, having been described as "more of an art than a science"[15].

If taking the semiclassical limit will transform a quantum system into a classical one, then here's the obvious question for a chaos theorist: are there quantum dynamical systems which, when you take the semiclassical limit, transform into *chaotic* classical dynamical systems? What characterises these? Are there any features of a quantum system which betray the fact that its underlying system will be chaotic?

Such quantum systems indeed exist and they're characterised by certain statistical trends in their spectra. Remember, a quantum dynamical system will have a spectrum of energy levels and this sequence of values can be subjected to statistical analysis.

Analysing such spectra, looking for "signatures" of (underlying) chaos is exactly the sort of thing quantum chaology is concerned with. The surprising truth of the matter, though, is that when the underlying classical dynamics are *not* chaotic, the quantum system's spectrum appears to be random. When the underlying classical dynamics *are* chaotic, the spectrum looks non-random (in statistical terminology, it "displays correlation" – that is, there's evidence that the various eigenvalues are related to each other according to some mathematical scheme). This reverses what we might have expected, our superficial impressions of "chaos" and "randomness" tending to be adjacent.

Gutzwiller's Trace Formula is an important tool in semiclassical physics because it shows a relationship between properties of a quantum dynamical system and those of its underlying classical dynamical system (remember, that's the system you get if you let the Planck constant shrink down to zero). More specifically, it relates the spectrum of energy levels of the quantum system (one side of the equation) to the lengths of the periodic orbits of the classical system which underlies it (other side of the equation). Recall that the Selberg Trace Formula is just one particular case of the GTF, where the classical system involves geodesic motion on a compact Riemann surface, the GTF being applicable in a much wider context. In other words, the GTF "generalises" the STF. When Martin Gutzwiller published his formula in 1971, there was no chaos theory as such and certainly no quantum chaology, but his discovery was to become a significant feature in the body of work that would eventually be known by that name.

Consider the (pure mathematical) spectrum of eigenvalues associated with a compact Riemann surface. This is what one side of the STF concerns and what we can think of roughly as "the sound the surface would produce if you could make one and hit it". This (entirely non-physical) spectrum corresponds to the spectrum of *energy levels* of a quantum dynamical system in the GTF.

The resemblance between the Riemann–Weil Explicit Formula and the STF discussed in Chapter 28 has been noted to extend to a very close resemblance between the RWEF and the trace formula of Gutzwiller[16]. It has also been shown how the RWEF can be interpreted as a trace formula on a "suitable space" [17]. This brings forth the hazy possibility that the RWEF might *also* be something like a special case of Gutzwiller's formula. Although there are immediate mathematical problems with this suggestion, we can at least entertain the image of a chaotic dynamical system where a Gutzwiller-like trace formula relating the system's periodic orbits to the energy levels of its quantum counterpart would reduce to a Riemann–Weil-like explicit formula relating primes and zeros. This dynamical system wouldn't have to be something which could physically exist – it could just be an abstract mathematical entity. If that seems puzzling, think of it like this: A number theorist looking at Gutzwiller's formula for the first time could potentially notice its resemblance to the familiar RWEF, and might then try to build on this by starting "*Suppose we define a dynamical system as follows…*" without any concern for physical reality. The aim would be to contrive a mathematical description of a dynamical system (using number theoretical tools) such that when the GTF was applied, it would reduce down to the RWEF.

The resemblance between the GTF and the RWEF suggests that quantum chaology could be relevant to the spectral nature of the Riemann zeta zeros. Thirteen years after Gutzwiller published his formula, and four years after he observed its similarity with Selberg's, an "intimately linked"[18] conjecture was put forward which connected quantum chaology and random matrix theory. In 1984, Oriol Bohigas, Marie-Joya Giannoni and Charles Schmit published a paper[19] showing how the relationship between the spectra of quantum dynamical systems and the chaos of their underlying classical systems could be understood using random matrix ensembles. Certain ensembles were shown to correspond to certain classes of classical dynamical systems. In each case, the spectra of the ensemble's matrices statistically resemble the spectra of the quantum dynamical systems which the relevant type of chaotic classical systems underlie.

As mentioned earlier in this chapter, GUE spectra in particular have become associated with quantum systems whose underlying classical mechanics are chaotic. Underlying those quantum dynamical systems whose spectra resemble GUE spectra are classical dynamical systems characterised not just by chaos, but also by their *lack of time reversibility*. Many simple dynamical systems are *time-reversible*. This means that if we were to (hypothetically!) film the system in motion and then play the film backwards, it would still show a possible motion of the system – what we'd see would still adhere to the rules governing the system in its usual "forward time" context. Examples of systems without this property (time-irreversible systems) are easily described in classical mechanics if we introduce friction, an immediate example being a simple pendulum. By watching the pendulum-on-film to see if it gained or lost energy, it would be possible to deduce which way the film was running. Examples in an idealised frictionless setting are less obvious, but relatively simple ones can be produced involving charged particles moving in magnetic fields [20].

The various other random matrix ensembles similarly allow quantum chaologists to "fingerprint" quantum systems via their energy spectra, identifying the *class* of classical systems underlying them (chaotic or not? time-reversible or not?).

WHAT'S DOING THE VIBRATING?

We return to the question: if the Riemann zeta zeros correspond to the spectrum of something, what can we deduce about that "something"? Well, it now seems to resemble a quantum dynamical system with underlying classical dynamics that are chaotic and time-irreversible. Despite the overwhelming evidence, nothing has been proved, but the nontrivial zeros of the Riemann zeta function do look strikingly like the energy spectrum of this kind of quantum mechanical system. No one particular system has yet been found with an exactly matching spectrum, but all the statistical fingerprints are so clearly present that any remaining doubts have been dispelled.

Here's a quick account of how this came to be known: Michael Berry was working on quantum chaology in Bristol in the 1980s. During that time, Andrew Odlyzko was busy computing Riemann zeta zeros in New Jersey, producing huge amounts of evidence in favour of Montgomery's GUE Hypothesis (this roughly stating that the zeta zeros are "statistically identical" to the spectrum of an infinitely large random matrix from the GUE class). In 1984, Bohigas, Giannoni and Schmit introduced random matrix theory to quantum chaology. Berry naturally became acquainted with this innovation. Then, having read Don Zagier's popular article on the distribution of primes (the one I quoted from a few times in Volumes 1 and 2)[21] and become interested in the zeta zeros, he eventually came to learn about Montgomery's GUE Hypothesis.

Berry saw early circulating versions of Odlyzko's calculations (which weren't published until 1987) and in 1986 published the landmark paper "Riemann's zeta function: A model for quantum chaos?"[22]. He acknowledged some related earlier work by Boris Pavlov and Ludvig Faddeev (1972)[23], and Martin Gutzwiller (1983)[24] which related the zeta function to aspects of quantum physics, but Berry's paper crossed into entirely new territory. Backed up by a significant body of evidence in the form of Odlyzko's calculations, it suggested for the first time the possible existence of an actual (chaotic and time-irreversible) classical dynamical system underlying a quantum system whose spectrum of energy levels would exactly match the Riemann zeta zeros.

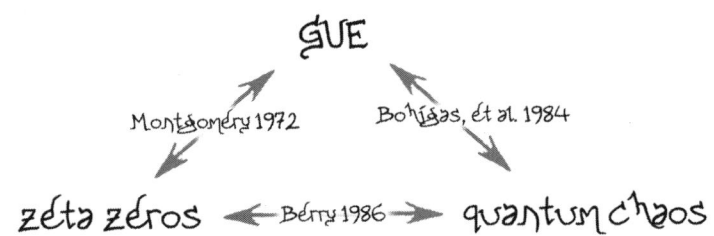

The paper, in other words, made the extraordinary suggestion that the Riemann zeros (the source of the frequencies of the spiral waves underlying the number system) could be somehow rooted in quantum chaos. And it went further, specifying a number of very specific properties which the implied dynamical system must have and discussing various promising possibilities and likely obstacles involved in seeking it.

Significantly, this work involved an explanation of how certain deviations from "pure GUE" behaviour in the Riemann zeros (originally noted as unfortunate "discrepancies" in Odlyzko's calculations – see page 37) could be explained simply and beautifully in terms of semiclassical physics[25]. This leads to the conclusion that the random matrix theory approach to the spectral nature of the Riemann zeros will only gets us so far and that beyond that (to get the fine detail) we must look to quantum chaology.

Berry, together with Jonathan Keating, elaborated on this theme in the years that followed. In 1998, they published the paper "$H = xp$ and the Riemann zeros"[26] which made further progress towards identifying the required dynamical system but presented only a partial solution to the problem. A number of other partial solutions to this problem have emerged, many inspired by ideas from various areas of physics[27]. In 1999, Berry and Keating produced a thorough survey of published work which had applied quantum chaological thinking to the mathematical mysteries of the Riemann zeta function[28].

So, we're looking at the profoundly weird possibility that there could be a chaotic, time-irreversible, classical dynamical system which underlies a quantum mechanical system whose spectrum corresponds to the heights of the nontrivial Riemann zeta zeros.

What sort of thing are we talking about here?

Recall that Marcus du Sautoy wrote:

> "*We have all this evidence that the Riemann zeros are vibrations, but we don't know what's doing the vibrating.*"[29]

Although this is an exciting and thought-provoking assertion, it perhaps involves a subtle blurring of the truth. We have a large body of computational evidence (via random matrix theory) that the zeta zeros are *spectral* in nature, but this doesn't imply that they're a spectrum of *vibrational frequencies*. The bulk of the evidence in favour of the spectral interpretation suggests rather that they're a spectrum of *energy levels*. The only evidence for the interpretation of the zeros as a spectrum of frequencies is the resemblance of the Selberg Trace Formula and Riemann–Weil Explicit Formula explored in Chapter 28. But we shouldn't let that get in the way of du Sautoy's memorable aphorism, which I'd argue is basically true "in spirit" – recall that energy and frequency are directly linked in quantum physics via Planck's constant. And as a result of the work I've described in this chapter (mostly due to Michael Berry and his collaborators, predominantly Jon Keating) we now have some clues as to "*what's doing the vibrating*". However, rather than these clues in any way lessening the sense of mystery, the fact that quantum physics and chaos theory are now involved has greatly intensified it.

If the sort of dynamical system Berry and Keating have partially described were to exist, then this would pleasingly bring together our "Exhibit A" and "Exhibit B": Not only would the random matrix theory-based evidence be accounted for, but also the RWEF–STF resemblance, in the sense that the Riemann–Weil Explicit Formula, relating primes and zeta zeros, could be understood (like the Selberg Trace Formula) as a kind of trace formula for a chaotic dynamical system.

chapter 31
the world turned upside down

So, theoretical physicists who study matter at the scale of atoms and subatomic particles have been able to shed some light on the mathematically mysterious Riemann zeros. They've shown that the zeros collectively possess certain statistical "fingerprints" which suggest that they're the spectrum of energy levels, or frequencies, of something resembling a physical system.

I feel that it's important to emphasise just how astonishing this is.

The usual relationship between mathematics and physics involves mathematicians shedding light on issues of interest to physicists. Because physics is formulated in terms of mathematical models, it's not surprising that developments in mathematics can lead to an expansion of our understanding of the physical world.

Some people protest: "But mathematical structures show up all over physics, so what's so astonishing about the Riemann zeta function showing up in physics?"

To such people I must emphatically point out that it's the other way round!

We've been discussing a structure underlying the number system which looks like *one particular example* of a class of physical systems. The usual scenario would involve a physical system's behaviour corresponding to one particular example of a class of mathematical structures. There's something categorically different about this[1].

We should perhaps stop and consider once more how inarguably *there* the "spectrum" of zeta zeros is. As I repeatedly stressed in Volume 2, it's "built into the structure of reality", even though it was only discovered in 1859. Attempting to argue that it's nothing more than a psychological or cultural construct would be foolish or misguided. The only concession that can be made to such a suggestion would be to consider what Gérald Tenenbaum and Michel Mendès France have written...

> "As archetypes of our representation of the world, numbers form, in the strongest sense, part of ourselves, to such an extent that it can legitimately be asked whether the subject of study of arithmetic is not the human mind itself. From this a strange fascination arises: how can it be that these numbers, which lie so deeply within ourselves, also give rise to such formidable enigmas? Among all these mysteries, that of the prime numbers is undoubtedly the most ancient and most resistant." [2]

...and then to think of the "spectrum" of nontrivial zeta zeros as some very deep, archetypal imprint of consciousness, or something of that nature. But that's another matter (which we'll eventually return to). Those who argue that all of mathematics is merely a human creation are seeking to convince others that there is no genuine mystery or wonder to be found in it. Even if the view that the zeta spectrum is a reflection of some mental or psychic structure were somehow shown to be accurate, that would still be a sufficiently strange and unexpected phenomenon as to be worthy of intensive study. And this still wouldn't explain away the near undeniable link with quantum chaos.

The doctoral thesis of Nina Snaith [3], written under the supervision of Jonathan Keating (whose supervisor was, in turn, Michael Berry), refers to "those in search of the elusive physical system lurking behind the Riemann zeta function". The choice of wording – "lurking behind" – is noteworthy.

A question looms before us: what, ultimately, are we dealing with here?

THE "RIEMANN OPERATOR"

Computationally (and otherwise), the evidence is now so overwhelming that the validity of the spectral interpretation is universally accepted. But although we have some clues, we still don't know "what's doing the vibrating". No one is denying the validity of Berry and Keating's work, so it's fair to say that whatever it is should have some kind of close relationship with quantum chaological systems of a particular type – they've even been able to provide a detailed list of precise characteristics which the system in question would have (if it were ever to be constructed or found).

The hypothetical system which has all of these characteristics and, crucially, whose spectrum matches the heights of the zeta zeros is known as the *Riemann dynamics*. But it is hypothetical – no one has been able to describe it precisely in mathematical terms, despite extensive efforts. We have knowledge about many of the properties it must have, but we're unable to find anything which has all of those properties.

In the process of introducing eigenvalues in Chapter 27, I also made reference to operators. Operators show up in maths and physics, sometimes representable by matrices, sometimes not, but very often with a spectrum of eigenvalues. The Riemann dynamics would be describable in terms of an operator – the spectrum of energy levels of the quantum dynamical system which it underlies would be the eigenvalues of that operator. Again, this is currently a hypothetical operator (the *Riemann operator* or *Hilbert–Pólya operator* mentioned briefly near the beginning of Chapter 28). But many of the mathematicians and physicists involved in this area of research feel sure that it must exist. Otherwise, why would the Riemann zeta zeros look so convincingly as if they had arisen as the spectrum of a very particular kind of operator?

In the same sense that we can say the Riemann zeros appear to be the eigenvalues of an infinitely large random matrix (accepting that this is not a mathematically precise assertion), we can say that they appear to be *the spectrum of a random operator*. But, paradoxically, if a "Riemann operator" does exist, it will be a unique, incredibly special operator (being built into the foundations of the number system in some deeply mysterious way) and in no sense "random".

Again, I must stress that the zeta zeros appear to correspond to not just a spectrum, but a *very particular type of spectrum*. It's puzzling, though, that something so archetypal (and presumably unique) as the set of zeta zeros should appear to correspond to one, seemingly arbitrary, member of an infinite class of things – as if it were not as special as we thought, but just one member of a wider population. But there's only one set of counting numbers (with its unique zeta function, zeros, *etc.*) so what's going on? We just don't know.

This is quite possibly the strangest thing currently known to humanity because we're discussing something which seems to "precede" the number system, something "lurking behind" the familiar structure with which we mediate reality.

THE MYSTERY OPERATOR AND THE RIEMANN HYPOTHESIS

If we were able to find or even just describe a physically plausible chaotic dynamical system and prove it to have all of the properties specified by Berry and Keating, then all of the nontrivial Riemann zeta zeros would necessarily lie on the critical line and so *the Riemann Hypothesis would be true*. Astonishingly, then, the quest for a proof of the Riemann Hypothesis has been closely linked to the quest for a very particular hypothetical chaotic dynamical system.

It works like this. The eigenvalues of an operator can be real or complex numbers. A simple mathematical manipulation can be used to rotate the complex plane 90 degrees anticlockwise and then shift it half a unit down, so the critical line ends up on top of the real axis.

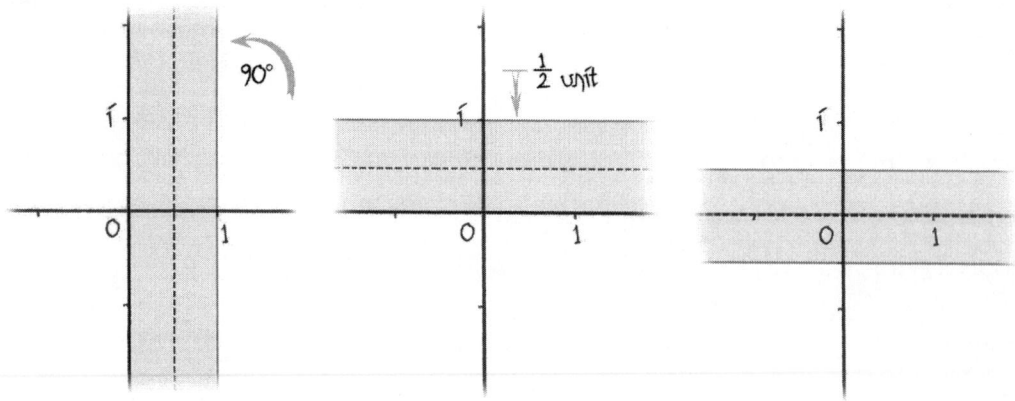

In this way, the question of whether the nontrivial zeros of the zeta function all lie on the critical line (that is, the RH) can be transformed into a question about whether all the (potentially complex) eigenvalues of a hypothetical operator are real numbers. With this correspondence, an eigenvalue of the operator which lies in the complex plane but not on the real axis would correspond to a nontrivial zero off the critical line.

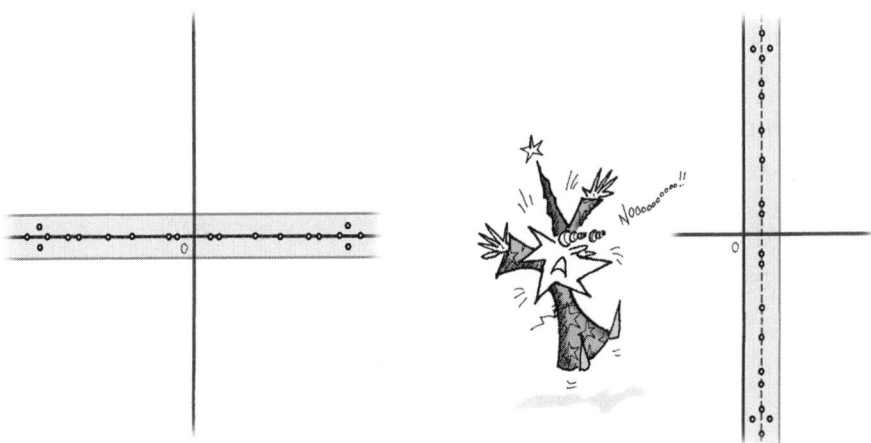

Berry and Keating's hypothetical operator belongs to a certain class I mentioned in Chapter 28, the class of *Hermitian* operators. Because of the way these are defined, their eigenvalues are always all real numbers. So, if a Hermitian operator can be found which satisfies all of the conditions specified by Berry and Keating, then the nontrivial Riemann zeta zeros are forced to lie on the critical line and so the RH is true. This goes right back to the original Hilbert–Pólya idea mentioned early in Chapter 28. At the time Hilbert and Pólya put the idea forward, though, there was no random matrix theory, Selberg Trace Formula, Riemann–Weil Explicit Formula, Gutzwiller Trace Formula, GUE Hypothesis, quantum chaology or Odlyzko-generated computational evidence, so their insight was prescient almost to the point of being prophetic.

THE GREATEST MYSTERY

The fact that there are spiral waves underlying the arrangement of primes was already remarkable. Recall Bombieri's words:

"To me, that the distribution of prime numbers can be so accurately represented in a harmonic analysis is absolutely amazing and incredibly beautiful. It tells of an arcane music and a secret harmony composed by the prime numbers." [4]

Now, on top of that *"absolutely amazing and incredibly beautiful"* situation, we have the extreme weirdness of these harmonics seeming to originate with some unknown vibrating thing "lurking behind" our number system.

In my decades of seeking out the strangest corners of reality, this would easily top the strangeness list. There's some physical-like dynamical system behind, within, or "at the root of" the number system! It has left its signature in the form of the zeros of the zeta function, these only being unearthed in the 19th century when Western mathematics had reached the necessary level of sophistication.

Despite having the appearance of an obscure issue on the outer regions of mathematics and physics, I'd argue that this matter of the "dynamical" origins of the Riemann zeros (which underlie the number system) is the great mystery of our age. Unlike the riddle of the Sphinx and Pyramids, Stonehenge, whatever it was the Knights Templar were getting up to, the "Roswell incident", the JFK assassination or the sudden, symmetrical collapse of New York's Solomon Brothers Building at near freefall speed in 2001, this mystery is not tied to time, space, historical contingency, culture, species or planet.

And yet it concerns the inner workings of what is widely thought of as "our" number system.

IMPLICATIONS

Beyond the truth of the RH, what might the implications be if someone were to identify the Riemann operator (or dynamics)?

"Berry is also convinced that there must be a particular chaotic system which [will] exactly duplicate the Riemann [zeros]. 'Finding this system could be the discovery of the century,' he says. It would become a model system for describing chaotic systems ...

It could play a fundamental role in describing all kinds of chaos. The search for this model system could be the holy grail of chaos…Berry believes the system is likely to be rather simple, and expects it to lead to totally new physics. It is a tantalising thought." (Julian Brown)[5]

Berry himself hasn't published anything (that I'm aware of) which speculates about "totally new physics", but here's a more sober excerpt from one of his papers concerning the hypothetical Riemann operator.

"[W]hen (if) the operator…is found, it will surely be simple, and will provide a paradigm for quantum chaology comparable with the harmonic oscillator for quantum non-chaology."

The *harmonic oscillator* is among the simplest and most important systems studied in quantum mechanics. Many complicated systems can be built from combinations of harmonic oscillators. (To get a vague idea of the relationship, think of the way complicated waveforms can be built from sine waves.) There's no comparable "fundamental building block" in quantum chaology. If one were to be identified, huge progress in that field could be made very quickly.

In concluding his book *Prime Obsession*, John Derbyshire hints at the enormity of what is at stake, but sensibly warns against idle speculation:

"And of course, if the physicists really do succeed in identifying a 'Riemann dynamics,' our understanding of the physical world will be transformed thereby.

Unfortunately, it is impossible to predict what things will follow from that transformation. Not even the cleverest people can make such predictions, and those who do should not be trusted." [6]

If the first sentence is correct, then the importance of the Riemann dynamics could end up being huge, affecting everyone, regardless of their interest in or knowledge of number theory. However, as Derbyshire suggests, little can be achieved by speculating. He goes on to describe Bertrand Russell's dismissive opinion of the significance of his own esoteric work on mathematical logic in the early 20th century. Russell, he argues, would never have believed that his work could have any practical application and yet he was paving the way for the development of computer technology, in particular for Alan Turing's cracking of the Nazis' ENIGMA code, and therefore shaping the entire history of the latter half of that century. The point here is that esoteric mathematical discoveries can have enormous and entirely unexpected effects. Derbyshire seems to imply that the "totally new physics" which Michael Berry supposedly predicted could give rise to a "totally new technology" and with it an entourage of unintended consequences.

In 1995, Alain Connes published a partial solution to the problem of finding the Riemann dynamics[7]. His highly original approach generated quite a lot of excitement in the mathematical community at the time. Almost two decades later, there's been no significant published progress in this direction and the excitement has died down considerably, but Connes' approach is still seen as a possible way forward for eventually proving the RH. The dynamical system he proposed had eigenvalues matching the heights of all Riemann zeros on the critical line, but unfortunately it didn't guarantee that there are no nontrivial zeros *off* the critical line.

By 2000, Connes' ideas had gone into widespread circulation, a key paper having been published in English (the 1995 paper was in French) and Berry was quoted as saying:

> "*I'm absolutely sure that if he's right, someone will find a clever way to make it in the lab. Then you'll get the Riemann zeros out just by observing its spectrum.*"[8]

Berry's description of someone "making it in the lab" seems compatible with the possibility of totally new physics and with it, perhaps, new technology. But what would such a thing *be*? What would it "mean"? Does the physical universe already contain any of these? If not, what would the consequences be of creating the first ever physical manifestation of such a (tremendously important) system?

It would be a type of oscillator, a physical system that exists at a range of possible energy levels (its eigenvalues). Because of the connection with chaos theory, such a (hypothetical) oscillator, wrote Julian Brown, could be "*the holy grail of chaos*", suggesting that to construct and study such a thing would, at the very least, lead to a monumental leap forward in our understanding of chaos theory.

To me, though, more important than the emergence of any new science or technology is how all of this might eventually change our perception of "reality".

The involvement of quantum chaology in the quest to prove the Riemann Hypothesis is often the source of considerable surprise for those who learn about it. It's been my experience that the better someone understands the details, the more surprising they find it. Like someone you know well behaving in a completely uncharacteristic way, these uncanny revelations about the nature of what we like to think of as "our" number system leave us bewildered, unsure of our assumptions about what we're dealing with. To put it as simply and clearly as possible, *the number system appears to be something radically different from what we thought it was.*

Grappling with the idea that something like (but not) a physical system could be lurking behind the number system, a fanciful image which came to mind some years ago was that of "peeling away the surface of reality", in a cartoon-like context, to find some hidden circuitry or extravagant clockwork mechanism hidden beneath it – some unexpected "understructure" to the world.

I've come across modern descriptions of traditional shamanism rather like this – using trance dancing, drumming, chanting, fasting or mind-altering substances to break through the surface of reality and experience the "circuitry" beneath. This leads us to the unlikely (but perhaps apt) image of current analytic number theory and quantum chaology as being a peculiar kind of 21st century Western shamanism. An important distinction here, though, is that traditional shamans served a sociological role within their communities (providing healing, finding lost objects, communicating with purported ancestral and spirit realms) whereas these particular explorations within mathematics and physics have no comparable role.

REALITY?

The idea of the world we inhabit being a simulation has recently become a theme in mass entertainment with the appearance of *The Matrix* trilogy (1999–2003), *The Truman Show* (1998), *Inception* (2010) and other such films. Bizarrely, in 2003, an article by physicist Paul Davies appeared in – of all places – the political commentary page of Britain's *Guardian* newspaper, using an imaginative but plausible argument involving "parallel universes" and some simple probability theory to propose that "this" reality is almost certainly a virtual reality simulation occurring in some "higher level" reality[9]. This was next to comment pieces about the potential value of psychotherapy in prisons, the low voter turnout in a recent election in northwest London and the funding of higher education. The article instantly grabbed my attention as I used to entertain myself with such notions as a young teenager, carrying out "thought experiments": What if the world is a simulation? What if I'm really just a brain in a jar, wired up to a machine? How can you prove to me that you're really real and not just a machine-generated projection? I used to challenge my parents on this point, much to their bemusement! I later found out about the philosophical concept called *solipsism*, read books by Philip K. Dick and saw *The Matrix* films and Richard Linklater's *Waking Life* (2001). It felt as if Western culture was beginning to think along similar lines.

If this world *is* a simulation (just thinking hypothetically) then the discovery that the number system appears to have a vibrational, sort-of-physical-but-not, "quantum chaological" foundation is, to my mind, the closest we've come to "hacking into the code" or witnessing anomalies which provide evidence that we're in something like a simulation. It's not my intention to argue that "this reality" – the one in which I'm writing this book (and you're reading it) – *is* a simulation, but these discoveries about the Riemann zeta zeros are, to me, of a similar order and quality of weirdness to that possibility.

During the final stages of preparing this volume, there was a burst of media interest in an obscure "preprint" article by three physicists based at Bonn University called "Constraints on the universe as a numerical simulation"[10]. A lot of the coverage misled casual readers into thinking that these researchers had uncovered evidence of the universe behaving as if it were a computer simulation. Typically, the cover of the March 2013 issue of the BBC science and technology magazine *Focus* boldly proclaimed:

THE UNIVERSE YOU LIVE IN IS A
HOLOGRAM
Reality may be stranger than you could possibly imagine
PLUS Why everything could be part of a computer simulation…including you.

Although the hologram reference wasn't connected to the preprint from Bonn, these topics were linked by the way they were presented on the cover (along with an inset image of a character from *The Matrix*!). And having sought out the preprint, I found that the abstract begins "*Observable consequences of the hypothesis that the observed universe is a numerical simulation performed on a cubic space-time lattice or grid are explored.*" In other words, the idea behind the paper is just a thought experiment: *If* the universe were a computer simulation, what might the observable consequences of this be (from the perspective of subatomic physics)? This is a lot less exciting than someone claiming to have observed such consequences, but even the fact that this possibility is being explored is a culturally interesting phenomenon.

Chapter 32
number and physics

We've just seen that, via the spectral interpretation of the Riemann zeta zeros, mathematics and physics are relating to each other in a most unexpected way. It's significant that this involves a link between *number theory* and physics. Mathematicians and physicists both often express surprise at the existence of this number theory/physics interface since number theory is the last branch of maths they expect to find showing up in connection with physics. Why is this? Let's quickly look at number theory's place within mathematics as a whole.

Put simply, number theory is that part of mathematics which concerns the counting numbers and their interrelations. We've encountered *analytic number theory* (Chapter 18) which also deals with functions on the complex plane and other such things, not just counting numbers, but this is just an indirect route to better understand the workings of the system of positive integers.

Number theory is arguably "there" as soon as you introduce counting. In order to pose some of the basic questions, no significant concepts are needed beyond addition and multiplication (which brings with it the notion of divisibility).

Traditionally, mathematics has been separated into three main branches: algebra, geometry and analysis. Each of these can be further split into sub-branches: there are many different algebras, geometries and types of analysis (for example, Clifford algebra, differential geometry, *p*-adic analysis). If maths were a tree consisting of all these branches and sub-branches, number theory could perhaps be thought of as the trunk, although that oversimplifies matters somewhat. Number theory in some sense underlies all three main branches, but they also all feed into it. That is, all three branches rely on the system of counting numbers, but there also exists (1) an important area of study called *algebraic number theory*, (2) a fairly minor (but interesting) one called *geometric number theory* and, as we've seen, (3) analytic number theory, which gave us the Riemann zeta function.

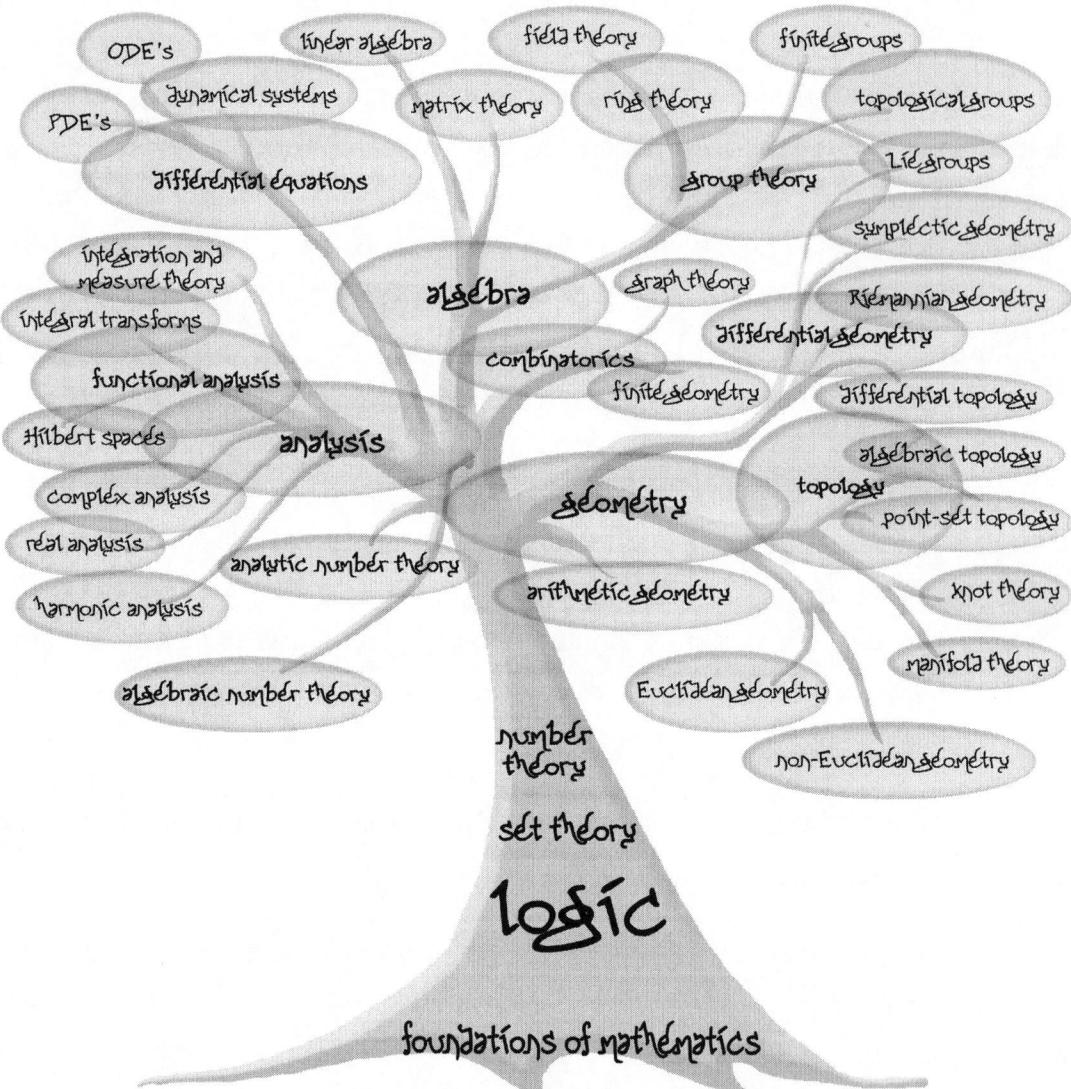

The tree analogy is helpful up to a point, but also problematic because every area of mathematics can be related to every other area, directly or indirectly (for example, we here see "topological groups" presented as part of group theory, but they're equally part of topology).

Number theory is also the most popular area of maths with amateurs (many of whom have no real interest in the other areas). This is because the counting numbers are so universal and accessible, while other areas deal with such things as *Hausdorff topologies*, *cyclic Fréchet modules* and *non-Abelian rings*, and it usually takes years of study just to understand the definitions of such things. Also, it's possible to easily state and understand problems like the Goldbach Conjecture (see Chapter 6). Although centuries of dedicated study by professional mathematicians have failed to resolve them, it's tempting for an amateur, when first encountering such a problem, to imagine that *they* might be the one to stumble on some solution which everyone else has overlooked.

Here are some typical examples of unproved conjectures from each of algebra, geometry and analysis[1]:

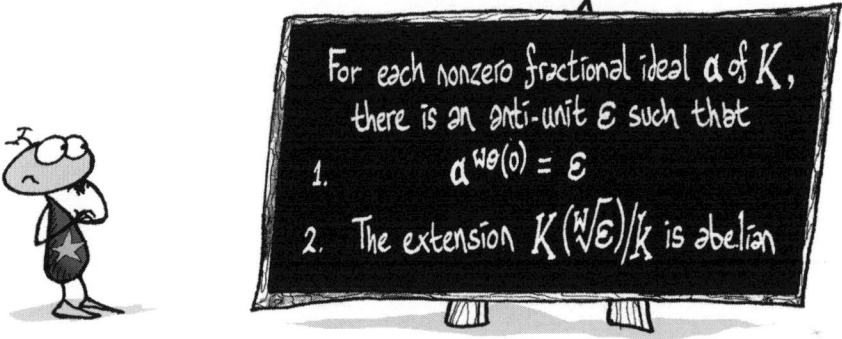

Let h be a non-negative increasing function on $[0, +\infty)$ and $\alpha > \frac{1}{2}$.
If
$$\int_0^1 \frac{h(tx)}{x}(1-x)^{n-1}dx \leqslant t^\alpha \quad \text{for all } t \in [0, +\infty),$$
then
$$\int_0^{+\infty} \frac{h(t)}{t}\frac{dt}{1+t^{2\alpha}} \leqslant \frac{\pi}{2}\prod_{k=1}^{n-1}\left(1+\frac{\alpha}{k}\right) = \frac{\pi}{2\alpha}\cdot\frac{1}{B(\alpha,n)}.$$

Here's a reminder of what the Goldbach Conjecture says:

for all $n \in \mathbb{N}$, there are primes p, q
such that $2n + 2 = p + q$

Here's another unresolved number theory problem, the *Twin Prime Conjecture*:

for all $n \in \mathbb{N}$, there is a prime $p > n$
such that $p + 2$ is also prime

Which would you attempt first?

As with any kind of difficult puzzle, some people can become fixated on pursuing solutions to such problems. This is usually with very little understanding of the background mathematics or awareness of what's known or has already been tried. It's a bit like attempting extreme mountain climbing with no previous experience, decent equipment or maps (just without the physical danger — although, as with all forms of obsessive behaviour, there are dangers in the realms of psychological well being).

Human interest in number theory goes back to at least Pythagoras and his followers (c. 500 BCE) who were intensely concerned with it (at least those simpler parts of it of which they were aware), and Euclid who made substantial contributions a couple of centuries later. But because the counting numbers have been known to humans for so long and thoughts in themselves leave no material record, certain problems could conceivably have been contemplated millennia earlier.

Paul Hoffman has written:

> "Of all branches of mathematics, number theory has traditionally been the most removed from physical reality. Seemingly abstract results in other esoteric areas of mathematics have been put to good use in physics, chemistry, and economics. This is not true of most results in number theory. If a proof of Goldbach's conjecture were found tomorrow, mathematicians would rejoice but physicists and chemists would not know how to apply the result, if indeed it has any application. Consequently, the contemplation of prime numbers has been regarded as mathematics at its purest, mathematics unadulterated by application. A few centuries ago, this kind of purity earned number theory the appellation 'the queen of mathematics'." [2]

Although there have been almost no physical applications until very recently, research in number theory has, in some cases, led to the expansion of theoretical knowledge in other areas of mathematics [3].

The great number theorist Godfrey Hardy stated in a 1915 lecture on the prime numbers:

"The theory of Numbers has always been regarded as one of the most obviously useless branches of Pure Mathematics. The accusation is one against which there is no valid defence; and it is never more just than when directed against the parts of the theory which are more particularly concerned with primes. A science is said to be useful if its development tends to accentuate the existing inequalities in the distribution of wealth, or more directly promotes the destruction of human life. The theory of prime numbers satisfies no such criteria. Those who pursue it will, if they are wise, make no attempt to justify their interest in a subject so trivial and so remote, and will console themselves with the thought that the greatest mathematicians of all ages have found in it a mysterious attraction impossible to resist." [4]

Hardy would probably have revised his opinion had he lived to see certain developments in recent decades. The relevant issue, though, is the extent to which the theory of prime numbers is normally understood as being "removed from physical reality". Going back to the tree diagram on page 66, it's the outermost "twigs" that generally make up the content of "applied mathematics", that is, mathematics put to use in tackling problems concerning physical reality. Number theory, down in the "trunk", has traditionally been understood as a core part of "pure mathematics", being of solely intellectual, rather than practical, interest.

Issues surrounding the RH are about as far away as you can get from the applied side of mathematics – it's at the heart of the purest of the pure. Carl Gauss famously described number theory as the "Queen of Mathematics" and mathematics as the "Queen of the Sciences" [5]. If there were to be a "Queen of Number Theory", it would have to be the study of Riemann's zeta function. Certainly, no one disputes the centrality of questions concerning the primes (particularly the Riemann Hypothesis) within number theory.

The sort of mathematics which physicists have become used to applying (and seeing show up after experimental data is analysed) involves number systems and structures far more sophisticated than the humble counting numbers – various types of mathematical spaces, often with four or more dimensions (sometimes infinite-dimensional) modelling physical phenomena, exotic geometries used to describe forces and systems, and even more abstract structures, yet further removed from our mundane experiences of number and space. In physics, we now expect to find these more complicated mathematical structures (things like vector fields, matrix algebras, fibre bundles and systems of differential equations) which all ultimately rest on the conceptual foundation of the number system. What we don't expect to find is number theory itself.

This is why, accompanying the appearance of number theory in this context, we often find expressions of surprise and astonishment. For example, in 1990, Jon Keating wrote a relatively non-technical article called "Physics and the Queen of Mathematics" in which he mentions the connections between quantum chaos and the Riemann zeta zeros, concluding:

> "*That there exist such similarities between the studies of the semiclassical limit of quantum mechanics and the theory of numbers is truly surprising. Even more remarkably, similar stories can be told for many other areas of theoretical physics. Indeed there are now conferences with titles like 'Number theory and physics'.*"[6]

Shortly, we'll be looking at some of these other areas of theoretical physics where number theory is starting to appear. What's important to note here is the common tendency for mathematicians and physicists to express *surprise* about this matter. There are a few fairly obscure instances of *additive number theory* being applied in nuclear physics and statistical mechanics, going back to the 1920s[7]. These would have probably seemed somewhat surprising at the time, but the quantum chaological connections made in the 1980s are something altogether different, for they involve certain very specific issues from mathematical physics being found to be unexpectedly relevant to questions of prime number theory.

The harmonic decomposition of the distribution of primes in terms of the zeta zeros is initially surprising, but once we've acknowledged that the zeta function can be defined in terms the primes, this surprise is somewhat lessened. This is the "circularity" described in Chapter 21 – the primes and the zeta zeros are somehow "dual" or interdependent. A much deeper layer of surprise comes from the almost certain validity of the spectral interpretation – the fact that the zeta zeros correspond to the spectrum of an unknown "something". But the seeming involvement of quantum chaos in this pushes the surprise up to another level. We have some of the deepest known physics (quantum chaology) shedding light on one of the most elementary issues in mathematics – the seeming lack of pattern in the prime numbers, something young children can understand.

I should stress, though, that the mathematics surrounding the primes (number theory) is *not* the mathematics which underlies quantum chaology. That, drawing on the mathematics involved in both quantum physics and dynamical systems theory, is very far removed from number theory indeed. Rather, as I've explained, it's the other way round: the physics (quantum chaology) seems to, in a way no one could ever have expected, "underlie" the distribution of prime numbers. However you interpret it, a significant link has been established between number theory and quantum chaology, and no one was expecting *that*.

OTHER AREAS OF PHYSICS

As Jon Keating mentioned in his 1990 article, beyond the quantum chaology connection, the Riemann zeta function and related mathematical structures have recently shown up in quite a few different areas of physics, much to everyone's surprise. This contributes to the feeling expressed earlier that *the number system seems to be something very different from what we previously thought*.

Concepts and results from analytic number theory have been shown, for no clear reason we can yet see, to have parallels in physics. I must reiterate that of all areas of mathematics, the contemplation of the subtleties of the prime distribution is about as far as it's possible to get from anything physically applicable, so these connections between number theory and physics came as a real surprise when they emerged in the 20th century.

In the early 21st century, we now find articles appearing in physics journals and online "preprint" archives where connections are being made between the Riemann zeta function and aspects of thermodynamics, signal processing, quantum mechanics (in contexts other than quantum chaology), string theory, cosmology and dynamical systems theory.

In the case of string theory and quantum cosmology, a powerful technique known as *zeta function regularisation* is now in widespread use. This can be traced back to an innovation in a 1977 physics paper by Stephen Hawking, where the behaviour of the Riemann zeta function is cleverly exploited in order to deal with the problem of "divergences", that is, problematic situations where a physics formula produces an infinite output[8]. The technique involves a new kind of zeta function based on Riemann's, now known as the *Hawking zeta function*. Other types such as *Hurwitz* and *Epstein zeta functions* are now used as part of a variety of specialised regularisation techniques.

Some of this physics-related or physics-inspired work is closely linked to recent approaches to the Riemann Hypothesis. These can be roughly separated into two historical currents:

The first we have already seen. It began with George Pólya's suggestion to Landau which gave rise to the spectral interpretation of the zeta zeros. It went on to involve the resemblance between the Selberg Trace Formula and the Riemann–Weil Explicit Formula, random matrix theory as developed by Wigner and Dyson, Montgomery's GUE Hypothesis, the Gutzwiller Trace Formula and the application of RMT to problems of quantum chaology, all of which eventually led to Michael Berry's suggestion of a (chaotic and time-irreversible) "Riemann dynamics". This is all rooted in the historical emergence of quantum mechanics. You might recall that

Pólya had been studying early quantum theory before working with Landau, this informing his suggestion which led to the spectral interpretation.

In the second current of research, the zeta zeros are given a very different physical interpretation. This interpretation is presumably related to the spectral one somehow (although a scheme has yet to appear which clearly relates them). Curiously, this current can also be traced back to George Pólya. After making his suggestion to Landau, Pólya went on to publish a number of papers on "locations of zeros" in the context of analytic number theory. In 1951, when Mark Kac (who we met in Chapter 22) was looking at a conjecture being put forward by the physicists Tsung-Dao Lee and Chen-Ning Yang, one of Pólya's theorems in this area came to mind. He suggested it to them and it went on to become a key ingredient in the main *Lee–Yang theorem*[9], a landmark result in the branch of physics known as *statistical mechanics.*

Statistical mechanics deals with gases and similar systems where the constituent particles or individual elements are too numerous to deal with in any "deterministic" mathematical way (imagine trying to model a huge billiard table with millions of balls involved). The systems are treated in a way which is analogous to how human populations are treated by sociologists looking at social trends – a statistical approach must be taken, as it's impossible to consider each individual separately.

The physical systems which statistical mechanics deals with have associated with them what's called a *partition function*, defined on the complex number plane $\mathbb{C}$. The partition function is the main object of study for a system in statistical mechanics, somehow "encoding" or encapsulating all of the important information about it. Zeros of this partition function correspond to physical processes called *phase transitions*. The Lee–Yang theorems are concerned with showing that for certain classes of physical systems, the zeros of their partition functions must lie on certain circles and lines in the complex plane.

In 1990, Bernard Julia published an article in which he presented a non-physical (but physics-inspired) statistical mechanics system constructed from the prime numbers[10]. He named this the *free Riemann gas* and applied the methods of statistical mechanics to study its properties. Already, this is quite a novel mixture of mathematics and physics. Julia calculated the system's partition function, which turns out to be none other than the Riemann zeta function. He then observed the similarity between the Lee–Yang theorems and the RH. Like certain Lee–Yang theorems, the RH seeks to restrict the zeros of a partition function (which just happens to be the Riemann zeta function in this case) to a vertical line in $\mathbb{C}$. This suggests that the Riemann zeta zeros correspond not only to energy levels of an unknown quantum mechanical system, *but also to phase transitions in some physical-like context*.

One interesting consequence of Julia's approach is that the behaviour of the Riemann zeta function at the point 1 can be understood in terms of a physical phenomenon called a *Hagedorn catastrophe*[11] (1 is the *pole* of the zeta function, where it's not defined, you might recall). Julia also pointed out that the "functional equation" of the zeta function (see Chapter 19) is related to a statistical mechanics phenomenon called *Kramers–Wannier duality*[12], and he went on to provide number theoretic interpretations for "ladders of fermion models" and other such physical phenomena. He acknowledged in his paper that he'd basically rediscovered (but also expanded on) something published some years earlier in a book by George Mackey[13] and that Donald Spector had also been working on something very similar around that time[14].

In 1995, Alain Connes and Jean-Benoît Bost took up this idea of the zeta function being a partition function and applied it in the context of *quantum* statistical mechanics. In this setting, the behaviour of the zeta function at the pole can be understood in terms of a physical phenomenon called *spontaneous symmetry breaking*. This paper led on to further work by Connes wherein he proposed the dynamical system mentioned in the previous chapter which *nearly* fulfilled the necessary criteria for the mysterious and much sought after Riemann dynamics (and so "nearly proved" the Riemann Hypothesis).

The interpretation of the Riemann zeta function as a partition function of a physical system goes further than this. Andreas Knauf and a number of colleagues have published extensively on *spin chains* (a class of abstract statistical mechanical systems) whose partition functions involve the zeta function in various ways [15]. Also, Daniel Fivel has independently interpreted the zeta function as a partition function in an entirely different context, that of *quantum entanglement* [16].

Donald Spector's work in this area involves the physical phenomenon known as *supersymmetry*. You might recall the *Mertens function* we met in Chapter 24 which involves looking at whether the number of prime factors in the factorisation of a given integer is odd or even. Spector related this function to something known to quantum physicists as the *Pauli exclusion principle* [17]. Another important feature of quantum mechanics, the Heisenberg uncertainty principle (see p.42), has been shown to parallel certain number theoretical phenomena by M.J. Shai Haran in his 2001 book *Mysteries of the Real Prime* [18].

As part of Berry and Keating's ongoing work, they have brought to light aspects of quantum mechanics corresponding to (1) the *Riemann–Siegel formula* (which is used to calculate Riemann zeros), (2) some number theoretical conjectures of Hardy and Littlewood [19], and (3) the Prime Number Theorem. The PNT was given other "quasi-physical" interpretations by Bill Parry and Mark Pollicott in 1983 [20] and (indirectly) by John Hannay and Alfredo Ozorio de Almeida in 1984 [21].

Further speculative work has led to plausible links being suggested between number theory and the following (all of which originate in physics): *diffusion processes, Bose–Einstein condensates, Brownian motion*, the *Fokker–Planck equation, 1/f noise, phase locking, pion-nucleon scattering*, the *Wiener–Khintchine duality relation, entropy*-related concepts, *Talbot effects* in classical optics, as well as scattering theory, aspects of string theory, quantum cosmology and quantum field theory [22].

A few of the promising approaches to the Riemann Hypothesis in recent circulation (those of Christopher Deninger, Alain Connes and Michel Lapidus) involve dynamical systems thinking: "flows" on certain spaces, the validity of certain (dynamical) trace formulas, *etc.* Other links have been forged between number theory and dynamical systems theory where entirely abstract, non-physical "systems" are defined in terms of number theoretical information or phenomena and then studied using techniques originally developed for application to physical systems[23].

Literature suggesting novel, physics-inspired approaches to number theoretical matters is now becoming fairly common. For example, in "Wave packets can factorize numbers" by Holger Mack *et al.*, published in an Italian physics journal in 2002, the authors explain:

> "*We draw attention to various aspects of number theory emerging in the time evolution of elementary quantum systems with quadratic phases. Such model systems can be realized in actual experiments. Our analysis paves the way to a new, promising and effective method to factorize numbers.*"[24]

I've taken a particular interest in this emerging interface between number theory and physics since I became aware of it in the late 1990s. Since 1999, I've been compiling and maintaining an extensive online archive on the subject[25].

Here are some more titles of papers of this kind. I've listed quite a few in order to give a general sense of the kinds of things which are being explored here. To help you see what's what, the number theory part of the title appears in italics and the physics part in boldface:

"A *number-theoretic estimate* for the **Thomas–Fermi density**"
"**Entanglement distillation protocols** and *number theory*"
"*Factorizations* and **physical representations**"
"A *number theoretic* interpolation between **quantum and classical complexity classes**"
"On the **quantum density of states** and *partitioning an integer*"
"An **electrostatic** depiction of the validity of the *Riemann Hypothesis*"

"*Prime decomposition* and correlation measure of **finite quantum systems**"

"**Wave-particle complementarity** and *reciprocity of Gauss sums* on **Talbot effects**"

"**Quantum Hamiltonians** and *prime numbers*"

"Modified *Möbius inverse formula* and its applications in **physics**"

"*Riemann zeros* and **periodic orbit quantization by harmonic inversion**"

"**Phase transition** in the *number partitioning problem*"

"*Riemann Hypothesis* and **short distance fermionic Green's functions**"

"**Inverse scattering, the coupling constant spectrum,** and *the Riemann Hypothesis*"

"*Ramanujan–Fourier series*, the **Wiener–Khintchine formula** and the *distribution of prime pairs*"

"The **classical gases** in the Tsallis statistics using the *generalized Riemann zeta functions*"

"Toward a **dynamical model** for *prime numbers*"

"**Thermodynamic limit** in *number theory*: Riemann–Beurling **gases**"

"**Tachyon condensation and brane descent relations** in *p-adic* **string theory**"

"*p-Adic and adelic* **free relativistic particle**"

"Path integrals for a class of *p-adic* **Schrödinger equations**"

"**Scattering** on *p-adic and adelic* symmetric spaces"

"**Lorentz-invariant Hamiltonian** and *Riemann Hypothesis*"

"*The Riemann zeta function* and the **inverted harmonic oscillator**"

"*The Riemann zeros* and the **cyclic renormalization group**"

"*The Riemann zeta function* and **vacuum spectrum**"

"On the subtleties of *arithmetical* **quantum chaos**"

"*Cyclotomy and Ramanujan sums* in **quantum phase locking**"

"Curiosities of *arithmetic* **gases**"

"*Number theory*, **dynamical systems** and **statistical mechanics**"

"**Attractors of random dynamical systems** over *p-adic numbers* and a model of **noisy cognitive processes**"

"*The Riemann zeta function* applied to the **glassy systems and neural networks**"

"Hecke Algebras, Type III factors and **phase transitions with spontaneous symmetry breaking** in *number theory*"

"*Riemann's hypothesis* and some infinite set of **microscopic universes of the Einstein's type in the early period of the evolution of the universe**"

"The emerging role of *number theory* in exactly solvable models in **lattice statistical mechanics**"

"The **thermodynamic formalism** approach to *Selberg's zeta function for PSL(2,$\mathbb{Z}$)*"

"*Selberg trace formula, Ramanujan graphs* and some problems of **mathematical physics**"

"*Mersenne primes*, polygonal anomalies and **string theory classification**"

"*Selberg zeta function and trace formula* for the **BTZ black hole**"

"A **quantum field theoretical representation** of *Euler–Zagier sums*"

"**Holography principle** and *arithmetic* of algebraic curves"

"**Heat kernel** and *number theory* on NC-torus"

"Duality and the *modular group* in the **quantum Hall effect**"

"The *Explicit Formula* and a **propagator**"

"The *Riemann* **magneton** *of the primes*"

"The **phase transition** of the *number-theoretical* **spin chain**"

"The **phase transition** in statistical models defined on *Farey fractions*"

"**Statistical mechanics of 2+1 gravity** from *Riemann zeta function* and Alexander polynomial"

Since the early 1990s, several conferences have been held on this interrelation between number theory and various aspects of physics. In 2006, a new journal called *Communications in Number Theory and Physics* was launched, further reflecting (and accelerating) this trend.

PHYSICS AND THE RIEMANN HYPOTHESIS

How do these developments affect the status of the Riemann Hypothesis? Clearly not all of this physics-related or physics-inspired literature directly concerns the RH, but a lot of it does, and anything which involves number theory indirectly concerns the RH.

> "[The Riemann Hypothesis has] *no longer just analytic number theorists involved, but all mathematicians know about the problem, and many realize that they may have useful insights to offer. As far as I can see, a solution is as likely to come from a…mathematical physicist, as from a number theorist.*" (Roger Heath-Brown, quoted by Karl Sabbagh)[26]

That the next significant new idea is as likely to come from physics as anywhere else may be relevant to the various remarks made in Chapter 25 about our inability to prove the RH being due to a big gap in our mathematical understanding.

> *"Despite the stunning advances linking Riemann's zeta function to 20th century physics, no one is predicting an imminent proof of the Riemann hypothesis. Odlyzko's numerical experiments and evidence amassed by physicists have convinced everyone that a spectral interpretation of the zeta zeros is the way to go, but number theorists say they are at least one 'big idea' away from even the beginnings of a proof."* (Barry Cipra)[27]

The Riemann Hypothesis is currently the ultimate unsolved problem in number theory, arguably in the whole of mathematics. In recent years, some researchers have suggested that a proof of the RH might not be that far away, but even if one were to come along, it might not clear up the whole mystery we've been considering.

Bearing in mind all that we've seen thus far (the spectral interpretation, the link to quantum chaology and all the other diverse physics connections which call into question the nature of the relationship between physical and mathematical reality), I'd be surprised if the eventual act of proving or disproving the Riemann Hypothesis doesn't open up a new, deeper, more mysterious unsolved problem which lurks behind it. The following chapters will attempt to address this.

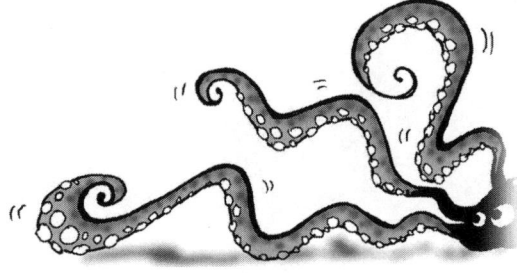

chapter 33

number and randomness

When applying the Prime Number Theorem, we're effectively dealing with the distribution of primes "statistically". The PNT tells us approximately how many primes we can expect to find in any given chunk of the number line. This can then be expressed in terms of *probabilities*. The probability of a "large, randomly selected counting number" being prime (putting aside for now the problems of precisely what is meant by "large" and "randomly selected") is approximately 1 divided by the natural logarithm of the number.

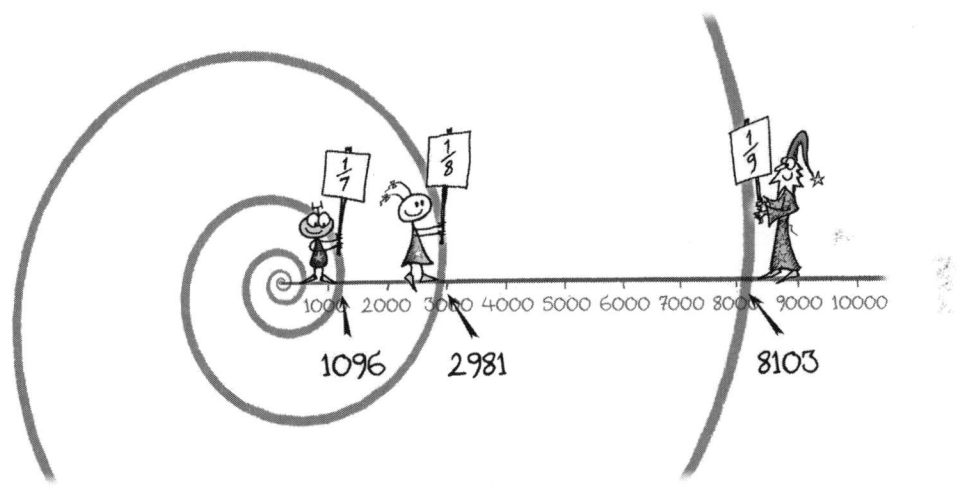

Recall that the natural logarithm can be related to the counting of coils of a "base-e" spiral.

What does this mean? If we were to put one white stone and nineteen black stones into a bag and then pick a stone at random, the probability that our chosen stone will be white is "1/20" or "1:20" or "1 in 20". This should be easy enough to understand. We're looking at the ratio of the *number of desired outcomes* (in this case the white stone being selected – there's only one possible way this can happen) to the *total number of possible outcomes* (in this case a single stone, black or white, being selected – there are twenty ways in which this can happen).

If we put two white stones and nineteen black stones in a bag, there are two desired outcomes (ways of picking a white stone) and twenty-one possible outcomes (ways of picking a stone of either type), so the probability is "2/21" or "2:21" or "2 in 21". The fraction version of this, "2/21", is the one we'll work with. We can express these fractional probabilities as decimals: 1/20 is 0.05 and 2/21 is 0.095238.... If we want to express them as percentages, we simply multiply by 100. So the chance of picking the white stone out of the twenty is $100 \times 0.05 = 5\%$, and the chance of picking one of the two white stones out of the twenty-one is $100 \times 0.095238...$ or about 9.5%.

Now, back to the prime numbers. I mentioned that the probability of a large, randomly chosen number being prime is approximately 1 divided by the logarithm of that number. If we name an actual counting number, say 2204332471, then *it's either prime or it isn't*, so this application of probability might seem confusing. But let's calculate it anyway and then consider what it means. Recall that the natural logarithm of this number can be found by drawing a circle of radius 2204332471 around the centre of what I've called the "e-spiral"[1] and then counting

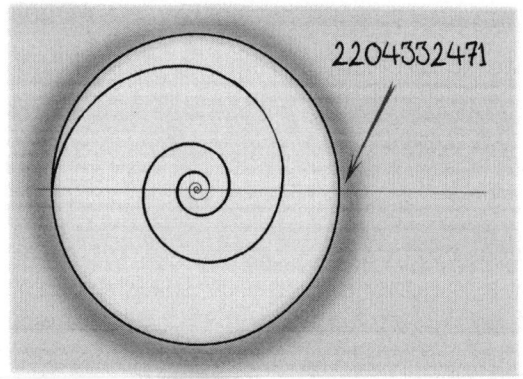

the number of coils between this circle and the disc of radius 1 with the same centre (far too small to be seen here). This "coil count" turns out to be approximately 21.513.

Dividing 1 by 21.513, we get approximately 0.04648.

This means that there's an approximately $100 \times 0.04648 = 4.648\%$ chance of the number $2\,204\,332\,471$ being prime. Note that mathematicians generally talk about a "probability of 0.04648" rather than a "4.648% chance", but these mean exactly the same thing.

But, as I've pointed out, $2\,204\,332\,471$ is either prime or it isn't. So what's this "4.648% chance" all about?

Well, the PNT tells us that the approximate number of primes up to the counting number $2\,204\,332\,471$ is that number divided by its natural logarithm, that is, $2\,204\,332\,471$ divided by $21.513...$, which comes out as about $102\,461\,861$. So this tells us that there are approximately $102\,461\,861$ primes to be found among the first $2\,204\,332\,471$ counting numbers. Now, imagine representing each of these as a white stone, the remaining $2\,101\,870\,610$ as black stones, and putting all $2\,204\,332\,471$ of these stones into a very large bag. What's the probability of picking a white stone (that is, a "prime")? It's the number of desired outcomes (the $102\,461\,861$ ways you could pick a white stone) divided by the number of possible outcomes (the $2\,204\,332\,471$ ways you could pick a single stone). Dividing, we get 0.0464. In other words, there's a 4.648% chance of picking a white stone. Equivalently, we can say that approximately 4.648% of the counting numbers up to $2\,204\,332\,471$ are primes. So, randomly choosing one of them, you have (approximately) a 4.648% chance of picking a prime.

Dividing 1 by a number's logarithm, then, we end up with a reasonably good approximate probability of that number being prime. But since the number is either prime or it isn't, it's not immediately obvious how to give the probability a precise mathematical interpretation. One way of looking at this would be to treat $2\,204\,332\,471$ as "0.04648 of a prime" (admittedly a rather peculiar idea) and similarly compute the probabilities for a long succession of counting numbers, keeping a running total of all of the resulting "pieces" of prime number. For any stretch of the number line, we could compare the actual number of primes encountered to this running total. We'd find that the more (and larger) counting numbers we consider, the more accurate this approximation becomes. This can be directly related back to the way the PNT works (the farther out you look, the more accurate the approximation will be).

THE CURIOUSLY FUNDAMENTAL ROLE OF PROBABILITY

Despite these explanations, looking at the primes in terms of probabilities may still seem confusing. A number is either prime or it isn't – these things don't change. The number system isn't like a weather system – it's just not the sort of context where you'd expect probability to be a relevant tool.

But it seems that however you look at it, "randomness" (despite the problems in defining this) is somehow involved in the distribution of prime numbers and therefore in the structure of the number system as a whole. As explained in Chapter 22, there's a branch of number theory called *probabilistic number theory* (having its origins in the 1930s) which exploits this randomness to obtain rigorous results. It's not particularly well known – in fact, many mathematicians are only vaguely aware of its existence. But once you're aware of it, it's hard to get away from the feeling that there's something quite strange going on here.

To understand what I mean by this, a little bit of historical perspective might be helpful. The theory of probability emerged in the 18th century when the mathematician Blaise Pascal collected, formalised and unified an extensive but disjointed repertoire of gamblers' tricks and techniques which had accumulated as a kind of specialised "folk knowledge" over centuries. For many people these days, the mention of probability conjures up images of bookmakers, horse races, lotteries, roulette wheels and casinos. Another place where it's regularly encountered is in weather forecasting ("a 20% chance of heavy showers"). Horse races and weather fronts are highly complex physical systems, with too many variables to predict exact outcomes, so probability is introduced as "the best we can do". So, it seems rather odd that probability theory should be so directly and naturally applicable to the *number system*, the rigid, timeless structure with which we order our reality.

Cambridge University mathematician Timothy Gowers touched on this matter when he wrote:

> "*It is interesting that the probabilistic model* [introduced to study various aspects of the primes] *is a model not of a physical phenomenon, but of another piece of mathematics. Although the prime numbers are rigidly determined, they somehow feel like experimental data.*"[2]

A similarly unexpected turn of events occurred in early 20th century physics when probability emerged as a fundamental ingredient in quantum mechanics. I must stress that this *isn't* about a level of physics where things get too complicated to predict so that you have to revert to probabilities as "the best you can do" (that exists as a branch of physics which was mentioned in the previous chapter – statistical mechanics). Rather, it's almost as if the probabilities produced by the mathematics involved are more fundamental than the particles that they're supposedly "about". As a result, it could be plausibly argued that the world is "built out of probabilities" rather than particles. Trying to write about this situation (rather than describe it mathematically) inevitably involves some vagueness and blurring of the truth – it's hard to grasp these ideas without a fairly extensive understanding of quantum mechanical theory, and that takes the form of highly elaborate mathematical models.

The fact that probability should show up like this in both number theory (the foundational level of mathematics[3]) and quantum physics (the foundational level of physics) suggests to me that our notions of randomness and probability may be somehow completely back-to-front, or inside-out. It *feels* as if (and I'm being intentionally vague here) randomness and probability are the underlying structure from which these systems emanate, rather than just concepts useful in the analysis of them.

Marcus du Sautoy here makes a point about the distinction between randomness in quantum mechanics and randomness in number theory:

"Prime numbers present mathematicians with one of the strangest tensions in their subject. On the one hand a number is either prime or it isn't. No flip of a coin will suddenly make a number divisible by some smaller number. Yet there is no denying that the list of primes looks like a randomly chosen sequence of numbers. Physicists have grown used to the idea that a quantum die decides the fate of the universe, randomly choosing at each throw where scientists will find matter. But it is something of an embarrassment to have to admit that these fundamental numbers on which mathematics is based appear to have been laid out by Nature flipping a coin, deciding at each toss the fate of each number. Randomness and chaos are anathema to the mathematician." [4]

When he writes of a "quantum die" deciding the fate of the universe, du Sautoy is referring to the fact that certain physical events at the quantum scale appear to be, at some very deep level, entirely random. Physicists are able to work with this in terms of their models and calculations, so they can more or less accept it. The idea has taken a while to settle in, though, and it encountered significant resistance when it first emerged. Einstein famously objected to (the then fairly recent) quantum theory with his famous remark *"God does not play dice with the universe"*. Alluding to this, the Australian number theorist Gerry Myerson once mischievously stated:

"God may not play dice with the universe, but something strange is going on with the prime numbers." [5]

There will more of this "strangeness" to come as we look further into this issue of randomness in the number system.

SO ARE THE PRIMES RANDOM OR NOT?

When first examining the distribution of primes in Volume 1, we considered the question of "order" or "pattern" in their arrangement. There were hints then about a "randomness" seemingly evident in the primes.

Having gone on to consider Riemann's formula for precisely determining the positions of the primes (in terms of his zeta zeros), it should be clear that any such talk of "randomness" must be carefully qualified, so we'll now re-evaluate the situation. Here's something quoted from Don Zagier back in Chapter 11:

> "*There are two facts about the distribution of prime numbers which I hope to convince you so overwhelmingly that they will be permanently engraved in your hearts. The first is that despite their simple definition and role as the building blocks of the natural numbers, the prime numbers … grow like weeds among the natural numbers, seeming to obey no other law than that of chance, and nobody can predict where the next one will sprout. The second fact is even more astonishing, for it states just the opposite: that the prime numbers exhibit stunning regularity, that there are laws governing their behaviour, and that they obey these laws with almost military precision.*" [6]

Zagier is presenting an almost paradoxical viewpoint here, two opposing facts which are both true: the primes are (or at least appear) thoroughly disorderly, the way weeds grow in amongst an orderly garden or lawn. At the same time, they're rigidly ordered (the quoted passage is taken from a lecture in which he introduces the Prime Number Theorem and then goes on to outline the role of the Riemann zeta zeros in the exact distribution of primes).

How can we reconcile these seemingly contradictory facts? In the first, Zagier seems to suggest that there is randomness inherent in the primes. In the second, he's talking about "*stunning regularity*", "*laws*" and "*military precision*" (we can assume that he's referring to both the PNT and Riemann's explicit formula).

So, is there a pattern, or *are* they random? This is the question people most commonly ask about the primes once they've grasped the definition and taken a look at a list of them. Don Zagier is saying that there *is* a pattern in the primes (if we take the obeying of laws "*with almost military precision*" to imply that some kind of "pattern" is involved) *and* that they are random.

WHAT EXACTLY *IS* RANDOMNESS?

Randomness is one of those phenomena which can appear to be reasonably straightforward on first consideration, but if you try to pin down exactly what it is, it can get rather confusing.

Have a look at these strings of digits:

372837283728372837283728372837283728372837

1010010001000010000010000001000000010000000

7430863506670357523541124346197239

3809525720106548586327886593615538127

Which of these would you describe as random?

The first two, I expect you'll agree, aren't random at all. There's a clear pattern evident in them which could be used to predict "what they'll do next".

The second two probably appear random to you.

The first of these seemingly random strings of digits *is* random, in the sense that it was produced by an electronic random number generator (a device involving a *noise diode* circuit).

The second, though, is an example of a sequence which is "statistically random" but not "truly random". You could run various statistical tests on it, each confirming that it does *seem* to be random, but in fact it's a string of digits taken directly from the decimal expansion of π. You might recall that π is an irrational number with an endless, non-repeating decimal expansion that starts

$$3.14159265358979323...$$

If you were to carry on until you had 1000 digits after the decimal point and then took the next thirty-eight digits, you'd get that fourth string shown opposite.

What this is meant to demonstrate is that whether or not something is random can depend on context. Before you were told that those digits came from π (unless you happened to recognise them!) they were effectively *random to you*. You had no way of knowing what they were going to do next. But once you learned that they were digits from the decimal expansion of π, they "ceased to be random to you". You could find a book or website containing an extended expansion of π and then determine exactly what the sequence does next. If you were taking part in some gambling activity involving these supposedly random digits, you'd then have a hugely unfair advantage over your competitors to whom the numbers *would* be effectively random.

In other words, randomness is (to some extent) relative to the awareness of the observer. In this regard, it's not an "absolute" concept. Rather than being something intrinsic, it can involve our relationship with the data under consideration.

> "*It is a mathematical contradiction to say that a series has no pattern; the most we can say is that it has no pattern that anyone is likely to look for. The concept of randomness bears meaning only in relation to the observer; if two observers habitually look for different kinds of pattern they are bound to disagree upon the series which they call random.*"
> (George Spencer-Brown) [7]

When discussing random matrix theory in Chapter 29, I explained that without a clearly defined context there cannot be any such thing as a "random matrix". Similarly, no individual number can (without a clearly defined context) be called a "random number". The numbers selected in a lottery draw are supposed to be random. Suppose the numbers 41, 7, 37, 11 and 28 are drawn. Does this mean that 41 can from now on be considered a "random number"? Of course not! 41 is no more inherently "random" than 42 – it's just that 41 was randomly selected *in this lottery*, whereas 42 wasn't. The point here is that any discussion of "random numbers" requires a context in which to be meaningful.

Interestingly, the definition of "randomness" is far from clear, even among probabilists and statisticians. Despite much effort having gone in to developing methods which can precisely distinguish between random and non-random series of digits, there's still no clear agreement – subtly differing definitions compete and are used in different contexts. Much has been written on the problem of defining randomness. Such discussions tend to be highly technical and so exclude anyone who isn't already familiar with the finer points of probability, statistics and computational theory. Still, it's a fascinating issue.

> "*It is evident that the primes are randomly distributed but, unfortunately, we don't know what 'random' means.*" (Robert C. Vaughan) [8]

Randomness is also dependent on something mentioned briefly in our discussion of random matrix theory, and in Chapter 22 when we met the best known example of one (the Gaussian or "bell curve") – a probability distribution. The relevant distribution determines the relative likelihoods of the various outcomes. So (as with random matrix theory) there are different "styles" of randomness, depending on the choice of underlying probability distribution. A simple illustration involves throwing two dice and adding the results. Although this is a random system with a random output, a total score of 7 is far more likely than a score of 12, as there are six ways in which we can obtain the former (1+6, 2+5, 3+4, 4+3, 5+2, 6+1) and only one way in which we can obtain the latter (6+6). A very simple probability distribution weights each of these outcomes accordingly.

As well as these considerations complicating any discussion of randomness (the existence of competing definitions, the importance of context and the dependence on a probability distribution), there are such concepts as *local randomness*, *global randomness*, *quasi-randomness* and *pseudorandomness*. Pseudorandom numbers are numbers generated by some mechanism which is deterministic and repeatable, but the output of which is complicated enough so as to be unpredictable and, ideally, statistically random. Pseudorandom numbers, then, are not "truly" random, but they're "random for all intents and purposes". They can be used to test hypotheses in statistical theory, *etc. just as if they were truly random*. Blocks of digits from the decimal expansion of π would be obvious examples. More generally, the production of pseudorandom numbers involves a "seed" and an algorithm. The seed is a small amount of data fed into the algorithm which then produces the pseudorandom output. Using the same seed with the same algorithm will always produce the same output, but by changing seeds we can endlessly generate blocks of numerical data which appear random.

One common definition of randomness which emerged from computational theory involves the notion of *compressibility*. Suppose that our purportedly random data is expressed as a binary string (a sequence of 0's and 1's). If it's possible to describe an algorithm (and seed) for generating this sequence which can be encoded into a binary string shorter than the original data string, then the sequence fails this particular randomness test (although it could still be considered pseudorandom if it passes appropriate statistical tests). A very simple example of compressibility would be the fact that to write "100 000 binary digits of π starting from the millionth digit of its binary expansion" takes up less space than those 100 000 digits themselves would, so despite appearing to be random for all intents and purposes, the string of digits would *not* be random according to the criterion of incompressibility. Similarly, you should be able to see how the second block of digits presented at the beginning of this section, if extended sufficiently, could be easily compressed.

The winning numbers in a lottery are sometimes chosen by subjecting numbered balls in a sealed chamber to vibrational forces. The physics involved follows deterministic laws, but is enormously complex. The output of this system (that is, the set of numbers which are selected) is so highly sensitive to the exact pattern of vibrations that it is entirely unpredictable. These lottery numbers are therefore considered to be truly random. Even if we set the system up so that it was indistinguishable from how it had been set up the first time and then attempted to repeat the process, we wouldn't be able to replicate the output. Exploitation of sensitivity in physical systems is the source of much true randomness (think of dice, coins and roulette wheels). Electronic devices for generating random output (such as the noise diode circuits mentioned earlier) rely on incredibly delicate fluctuations in physical processes which, despite being subject to well understood laws, are beyond the realms of predictability.

To generate random numbers, computers sometimes use high-precision clocks. These keep track of time (in seconds, typically) to many decimal places. If, at the exact moment that a program requires some random numbers, the sixth to tenth decimal digits (let's say) of the clock time are extracted, then these digits might as well have been chosen by something like a roulette wheel. Because some programs require random numbers to be generated at regular time intervals, there could potentially be problems with unintended and undesirable regularities showing up in the supposedly random output. To avoid this possibility, the digits are fed into a mathematical formula which thoroughly scrambles them. Although based on deterministic processes, the involvement of sensitivity means that this output would be considered "truly random" rather than "pseudorandom".

But perhaps to a being with an unimaginably vast intelligence, computational power and super-sensory perception, predicting the output of such systems could be effortless. Perhaps, to such a being, these sequences of numbers *wouldn't* be random. Interestingly, the question of whether randomness *exists at all* touches on theological questions, particularly the matter of free will and whether or not the universe is deterministic[9]. Various religious traditions have made use of *cleromancy*, a practice involving seemingly random processes, to reveal the supposed will of their gods, so something a secular observer might classify as random output could be understood as a divine channel of communication by someone else.

Arguably, the most "truly random" thing in the known universe is the timing of certain emissions from atomic nuclei (the phenomenon known as *radioactivity*). The inherent uncertainty of the quantum realm rules out the possibility of in any way predicting these emission times (regardless of vast intelligence or super-sensory perception). This can be exploited in order to generate random numbers. A detector (like a Geiger counter) is required, together with some radioactive substance that has an appropriately high rate of emission. The detector is connected to a computer which records the emission times. A simple and foolproof way to generate a sequence of random binary digits is to wait for four emissions, measuring the time between the first and second, and the time between the third and fourth. If the former is longer than the latter, that's interpreted as a "1", if it's shorter, that's a "0" (and if they're the same, we discard them and wait for the next four emissions)[10]. This is effectively like getting the universe to toss a coin rapidly and repeatedly. From these 0's and 1's, random numbers in any specified range can be straightforwardly constructed.

PARADOXICAL LANGUAGE AND POETIC LICENSE

Here's what a few learned authorities have had to say on the topic of randomness and prime numbers:

> "*Prime numbers have always fascinated mathematicians. They appear among the integers seemingly at random, and yet not quite: There seems to be some order or pattern, just a little below the surface, just a little out of reach.*" (Underwood Dudley)[11]

> "*The primes have tantalized mathematicians since the Greeks, because they appear to be somewhat randomly distributed but not completely so.*" (Timothy Gowers)[12]

> "*Primes are part of the scaffolding of the number system, but the structure of this scaffold is still obscure, still hidden despite the research of centuries. Prime numbers*

are imbedded throughout the system, yet no pattern in their apparently random appearance can be detected, no outward signs distinguish them from non-primes... They are an infinite code that no one has ever been able to crack." (Jane Muir)[13]

"*The primes appear to follow a kind of random distribution, and experiments with computers for large numbers of primes bear that out.*" (Enrico Bombieri)[14]

Note the choice of wording used here: "*appear among the integers seemingly at random*", "*appear to be somewhat randomly distributed*", "*their apparently random appearance*" and "*appear to follow a kind of random distribution*". All four authors are being careful not to say that there *is* randomness inherent in the primes, but that there *appears* to be.

This could be informed by some combination of the facts that: (1) the primes are fundamental to the number system and can't be any other way – they're just *like that*, (2) we now know about the role of the zeta function in precisely governing the distribution of primes, (3) despite being widely suspected, the truth of the Riemann Hypothesis (which is linked to the primes and, as we'll see, the issue of their "randomness") is not known, (4) as Bob Vaughan has pointed out, we're not quite sure what we mean by "random"!

Overriding all of this, however, is the fact that we can systematically (although inefficiently) compute the sequence with, for example, the Sieve of Eratosthenes, so they're at best *pseudo*random. It's fair to say, then, that these authors are using the word "random" in a fairly loose, colloquial sense (there will be more of this to come).

The first two quotations contain an element of the paradox we noted in the Zagier quote earlier: the primes are "random but not random". In the third, Jane Muir also implies something like this when she describes the primes in terms of "*scaffolding of the number system*" (which suggests something orderly).

All of the statements would have been even more apt prior to Riemann than they are now. No doubt, many generations of mathematicians who never lived to know of the zeta zeros felt a "tantalising" sense of something "a little below the surface" of the prime distribution, a "code" waiting to be cracked. The fact that the primes underlie the system of counting numbers, arguably the ultimate embodiment of "order", leads to the reasonable expectation that there should be some kind of order in the primes. To the generations of mathematicians past who conceived of God or some such creative force at work in the Cosmos, there may have been a (conscious or unconscious) tendency to suspect the presence of such an entity lurking behind the mysterious distribution of prime numbers. From such a perspective, it would have been difficult to accept a complete lack of law and/or order in the primes. And, representing the secular 21st century, we have du Sautoy writing:

> *"It seems paradoxical that the fundamental objects on which we build our order-filled world of mathematics should behave so wildly and unpredictably."*[15]

You might recall this passage which I quoted in Chapter 8 from Apostolos Doxiadis' novel *Uncle Petros and Goldbach's Conjecture*:

> *"The seeming absence of any ascertained organizing principle in the distribution of the succession of the primes had bedevilled mathematicians for centuries and given Number Theory much of its fascination. Here was a great mystery indeed, worthy of the most exalted intelligence: since the primes are the building blocks of the integers and the integers the basis of our logical understanding of the cosmos, how is it possible that their form is not determined by law? Why isn't 'divine geometry' apparent in their case?"* [16]

It's possible to detect a touch of underlying collective frustration ("bedevilment") in the mathematical community resulting from the paradoxical tension between the "they're just like that", "carved in stone" aspect of the primes and this seeming randomness.

Although these statements reveal something about the way mathematicians *feel* about these particular mathematical issues, there's arguably a slight blurring of the facts going on here. In the course of this trilogy, we gradually came to see how there *is* an "organising principle" in the distribution of primes, based around the Riemann zeta function and its zeros. But we then also saw what the distribution of Riemann zeros looks like – *they* appear to be random (apart from a logarithmic formula, as with the primes, describing how they scatter "on average"). So, in a sense, the quoted remarks are simplifications, but they convey the spirit of the situation. That is, even though we now know a lot about the zeta function (and even if we were to prove the Riemann Hypothesis), until some completely new way of looking at these things comes along, there remains something almost disturbingly arbitrary about the number system when you look at the (absolutely central) primes/zeros phenomenon. And the number system is the framework on which we build "our order-filled world", as du Sautoy points out. So what's going on?

Most of the material I've quoted in the chapter is taken from pieces written by mathematicians for a general audience, mathematicians inspired by the subject matter to use such poetic language as "engraved in your hearts", "grow like weeds", "bedevilled", "great mystery", "exalted intelligence", "cosmos", "divine", "wildly", *etc.* Admittedly, Doxiadis was writing a novel and, in general, a certain amount of poetic license is to be expected when writing about something this difficult to convey directly. But perhaps there's something more to it than that. Many other quotations about the primes, zeta function and related issues have strong poetic elements. Here's a good example we've already seen:

> *"Why do the primes achieve such a delicate balance between randomness and order? And if their patterns do encode the behaviour of quantum chaotic systems, what other jewels will we uncover when we dig deeper? Those who believe mathematics holds the key to the Universe might do well to ponder a question that goes back to the ancients: What secrets are locked within the primes?"* (Erica Klarreich) [17]

"Delicate balance between randomness and order" is a phrase many number theorists

would endorse, but these issues of pattern, order and randomness cannot easily be given precise mathematical formulations. We're moving away from mathematics and edging into the realm of human perception and experience, approaching the limits of what can be expressed (hence the need for poetic language). This is something we'll explore further in the final chapters.

ORDER OR PATTERN?

It could be argued that an *order* has been revealed in the primes, but no *pattern*. The order is the regime of zeta zeros, which arranges the primes in their precise locations ("ordering" them). This is the "military precision", the obeying of orders. But there's still a lack of *pattern* because of a knowledge problem – we can't ever know *all* of the zeta zeros, so we can never fully "see what's going on" with the distribution of primes. Beyond this, the quantum chaology connections remind us that *we don't really know where these zeta zeros come from*, so any pattern in the primes which they underlie remains thoroughly mysterious until that issue is resolved.

To summarise, we now know that there's an intricate structure underlying the distribution of prime numbers, but as the origins of this structure are presently obscure (in the literal sense that we can't clearly see them), we're only slightly less in the dark when it comes to how the primes are distributed. Once we've taken into consideration their tendency to thin out according to the simple logarithmic rule known as the PNT, they're still "random" to us (in the colloquial sense that the writers in the previous section were using the word), in a similar sense that a block of far-flung π digits would be "random" to most people. In both cases, there's an absence of apparent pattern in the sequence.

To everyone before Riemann's discovery of the zeta function and its nontrivial zeros in 1859, the primes would have appeared random in their arrangement on relatively small chunks of the number line. On a larger scale, there had been an awareness since the early 19th century of the "average thinning out" dictated by the PNT, but this was a *statistical* regularity, not the basis of a precise pattern.

Of course, using the Sieve of Eratosthenes, it has always been possible to systematically determine the sequence of primes, so in that sense they're *not* random. And after 1859, it could almost be said, the sequence of primes became "even less random", at least to that small sector of humanity familiar with Riemann's explicit formula and the nontrivial zeros of his zeta function. Thinking like this, we see that the prime numbers are not "random" in a given stretch of the number line to someone with either the means to apply the Sieve of Eratosthenes (or a similar method) far enough, or the ability to compute appropriate ranges of zeta zeros and apply Riemann's explicit formula.

Because of the inevitable physical and temporal constraints of being finite mortal creatures, though, there are upper limits on how far we will ever be able to apply the Sieve or compute the Riemann zeta zeros. For this reason, then, we must accept that beyond a certain point on the number line (which could even be roughly estimated) the primes are, and will forever remain, *random to everyone*.

THE RIEMANN HYPOTHESIS AND THE "RANDOMNESS" OF THE PRIMES

In contrast to all this vagueness, paradox and carefully qualified language, there are aspects of the prime numbers' collective behaviour which can be directly related to the mathematics of "truly random" phenomena (think of idealised dice, coins, *etc.*).

In Chapter 24, we saw how the Riemann Hypothesis can be reformulated in terms of a bound on the growth of the PNT error (this involved what I called "power curves" enclosing the deviation graph). We also saw how this RH-related bound on the growth of the PNT error (which is equivalent to the growth of the "primeness count deviation" – see Chapter 13) can be related to the bound on the growth of the "heads-tails deviation" produced by a repeatedly tossed fair coin.

A reformulation of the RH in terms of the *Mertens function* was also presented. You may recall how this graph...

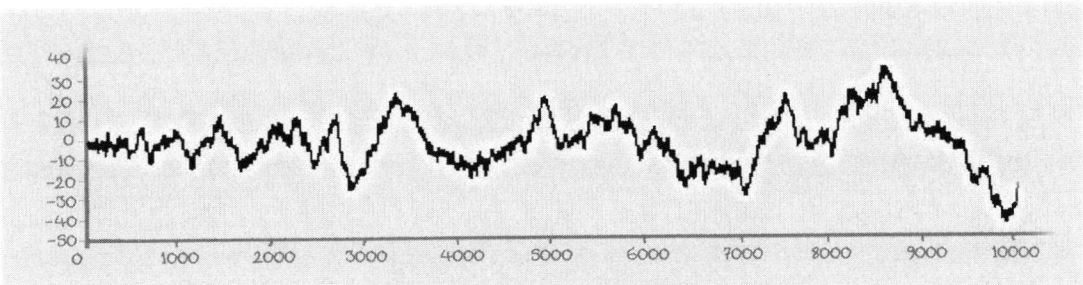

...was produced by looking at the prime factorisation of each counting number, considering whether (if there were no repeats) the number of factors was odd or even. This can also be used as a method to produce an infinite "heads/tails" sequence from the number system. The RH can then be shown to be equivalent to a necessary condition for that sequence to behave like a fair coin [18].

MAXIMAL RANDOMNESS

At a superficial level (and once their "logarithmic thinning out" is taken into account), the prime numbers resemble a randomly chosen sequence of counting numbers. But everything we've learned about the primes distinguishes them from other randomly chosen sequences which thin out at a similar rate.

Looming over this whole question of randomness-or-not is the inescapable fact that *the distribution of prime numbers cannot be any other way.* It's forced into the shape that it is so that the primes fulfil their role as multiplicative building blocks of the counting numbers. They're arranged in precisely the right way so that the number system can be built from them multiplicatively (in accordance with the Fundamental Theorem of Arithmetic) and the slightest change in their arrangement would completely undermine this.

They're a fundamental, universally accessible sequence etched into the structure of reality – they can't *really* be random. Yet in some ways they seem to be. Tenenbaum and Mendès France ask "*how a sequence so precisely determined as that of prime* numbers can incorporate so great a share of randomness"[19]. As we've seen, finding agreement on the meaning of the word "random" is problematic, but the conclusion which the mathematical community seems to have come to about this matter is that the primes, considering the very rigid constraints which they must satisfy, are "maximally random". They're as random as they can possibly be given the circumstances. The same authors attempt to explain:

> "One of the remarkable aspects of the distribution of prime numbers is their tendency to exhibit global regularity and local irregularity. The prime numbers behave like the 'ideal gases' which physicists are so fond of. Considered from an external point of view, the distribution is – in broad terms – deterministic, but as soon as we try to describe the situation at a given point, statistical fluctuations occur as in a game of chance where it is known that on average the heads will match the tails but where, at any one moment, the next throw cannot be predicted. Prime numbers try to occupy all the room available (meaning that they behave as randomly as possible), given that they need to be compatible with the drastic constraint imposed on them, namely to generate the ultra-regular sequence of integers."[20]

The fact that Mendès France and Tenenbaum mention "ideal gases" is noteworthy. They're not intending to suggest a direct link between number theory and the physics of gases, rather using something from physics as part of an analogy. But it's an apt analogy, as actual links between the primes and ideal gases *have* been made, by Bernard Julia and other researchers mentioned in the previous chapter. I mentioned this when discussing the branch of physics called statistical mechanics. Statistical mechanics involves randomness, so it connects physics and probability theory. Just as the seeming randomness in the primes makes possible the discipline of probabilistic number theory, it's the seeming randomness in the primes which makes the number system a possible object of study (albeit a novel one) for physicists who work in statistical mechanics. The mathematician Alain Connes has even used ideas from *quantum* statistical mechanics to construct a (pure mathematical) theory[21] which has enabled him to come as close as anyone to really making sense of the zeta function and its zeros (the maths involved uses *adeles*, an exotic breed of "supernumbers" which are closely tied to the primes [22]).

The primes "*try to occupy all the room available*", write Tenenbaum and Mendès France. Their words have been translated from French, so the intended meaning may be slightly distorted, but taken at face value they almost portray the number system (via the primes) as having a kind of *intention*. But what is this "occupy all the room available"? We can perhaps hazily picture it, but is it something we can precisely define? And "*as randomly as possible*"? We've seen that there are differing definitions of "randomness", that it depends on context – it's something hard to pin down. But if we adopt the notion of randomness widely used in probability theory, the one we'd associate with the issue of the fairness of a coin, then whether or not the primes behave as randomly as possible can be linked to whether or not the Riemann zeta function has zeros off the critical line.

Although these assertions about the primes being "maximally random given certain constraints" are not precise mathematical statements, the failure of the Riemann Hypothesis would imbalance the distribution of primes in such a way as to make them in some sense "less random" – there would be predictable "waves" of primes, with the sort of regularity exhibited by ocean tides. The closer we can guarantee all the nontrivial zeta zeros are to the critical line, the "more random" the primes become. The best we could possibly do in this regard is to show that they're all *on* the critical line, which results in the "maximal randomness" of the primes.

The RH hypothesises that all of the nontrivial Riemann zeta zeros lie on the critical line. We know that they all lie inside the critical strip. In Chapter 21, we encountered the idea of a "zero-free region" inside the strip and saw how the RH being true would imply that this region is as large as possible. This is equivalent to the primes behaving "as randomly as possible", in the sense of Tenenbaum and Mendès France. So, if the RH is false and some zeros stray off the line, then we have a narrower "zero-free region" and the primes will behave in a "less-than-maximally random" way.

We saw in Chapter 21 how the existence of a pair of zeros off the critical line would correspond to a spiral wave with rate of amplitude growth greater than 1/2. This amplitude growth would eventually produce huge fluctuations in the distribution of primes, which would undermine the optimal randomness promised by the RH. The farther off the critical line the zeros are, the faster these fluctuations grow and hence the more the optimal randomness is violated. So, insofar as we're prepared to talk about "randomness" in this way:

the extent of the randomness in the primes
> *is linked to*
>> the size of the zero-free region
>>> *is linked to*
>>>> the size of the error in the PNT

the RH being true
> *is linked to*
>> maximal randomness in the primes
>>> *is linked to*
>>>> maximal size of the zero-free region
>>>>> *is linked to*
>>>>>> minimal error in the PNT

RH being false
> *is linked to*
>> less-than-maximal randomness in the primes
>>> *is linked to*
>>>> less-than-maximal size of the zero-free region
>>>>> *is linked to*
>>>>>> less-than-minimal error in the PNT

AN AESTHETIC PERSPECTIVE

There's another consideration involving the issue of randomness in the primes (or, we could say, "disorder inherent in the number system") which is worth mentioning – an "aesthetic" one. Aesthetics don't come into mathematics directly, but outside the formal and rigorous framework of their published research, experienced mathematicians have been known to describe certain structures, theorems and proofs as "beautiful", "elegant", "gorgeous", *etc.* So there's some kind of loosely shared aesthetic in the mathematical community, but *this is not part of mathematics.* It's part of a separate discipline which involves elements of both psychology and mathematics. This discipline, which would be concerned with *our feelings about mathematics*, hasn't really been developed yet. There are a number of tentative writings which could be compiled, but that's about all.

We've already considered a reference to the "*delicate balance between randomness and order*" which the truth of the RH would imply was inherent in the primes:

> "*Proving the Riemann hypothesis won't end the story. It will prompt a sequence of even harder, more penetrating questions. Why do the primes achieve such a delicate balance between randomness and order?*" [23]

Paolo Ribenboim has suggested that this crucial balance between order and randomness can be appreciated from an artistic point of view:

> "[It is] *possible to predict with rather good accuracy the number of primes smaller than N (especially when N is large); on the other hand, the distribution of primes in short intervals shows a kind of built-in randomness. This combination of 'randomness' and 'predictability' yields at the same time an orderly arrangement and an element of surprise in the distribution of primes. According to Schroeder (1984), in his intriguing book* Number Theory in Science and Communication* [24], these are basic ingredients of works of art. Many mathematicians will readily agree that this topic has a great aesthetic appeal.*" [25]

In a similar vein, certain "aesthetic" arguments have been put forward for why the RH cannot be false. You might recall Bombieri having written:

"Even a single exception to Riemann's conjecture would have enormously strange consequences for the distribution of prime numbers. The primes appear to follow a kind of random distribution, and experiments with computers for large numbers of primes bear that out. If the RH turns out to be false, there will be huge oscillations in the distribution of primes." [26]

We saw that the extent to which any nontrivial zeta zeros stray from the critical line controls the extent to which the distribution of primes deviates from its average behaviour. If the RH fails, then we know that there are zeros off the critical line, these would cause "huge oscillations", the deviation would grow too fast and the primes wouldn't be maximally random. Bombieri went on to state:

"In an orchestra, that would be like one loud instrument that drowns out the others – an aesthetically distasteful situation… The failure of the Riemann hypothesis would create havoc in the distribution of prime numbers. This fact alone singles out the Riemann hypothesis as the main open question of prime number theory."

Note that the words "strange", "distasteful" and "havoc" have more to do with emotional and aesthetic reactions than anything which could be precisely formulated in mathematical terms.

Brian Conrey's survey article on the RH presents several points of evidence as to why it "should be" true, including the following:

"RH tells us that the primes are distributed in as nice a way as possible. If RH were false, there would be some strange irregularities in the distribution of primes…It seems unlikely that nature is that perverse!" [27]

Again, notice the use of imprecise, unmathematical language: "nice", "strange", "nature", "perverse". This is more about the interplay between psychology and mathematics than about mathematics itself.

We've encountered other mathematicians making aesthetic judgements concerning the possible truth or falsity of the RH:

"Mother Nature has such beautiful harmonies, so you couldn't say that something like [the Riemann Hypothesis] is false." (Henryk Iwaniec) [28]

"I would like the Riemann Hypothesis to be true, like any decent mathematician, because it's a thing of beauty, a thing of elegance..." (Aleksandar Ivić) [29]

"...Riemann's Hypothesis can be interpreted as an example of a general philosophy among mathematicians that, given a choice between an ugly world and an aesthetic one, Nature always chooses the latter." (Marcus du Sautoy) [30]

These three are basically saying that the Riemann Hypothesis must be true *because the reality we inhabit would be far too ugly if it weren't*. The beauty they're referring to corresponds to all of the nontrivial zeta zeros falling on the critical line and (equivalently) the "maximal randomness" of the primes – or, to put it another way, to the primes being as "well mixed" as arithmetically possible into the sequence of counting numbers.

THE ZETA ZEROS CONSIDERED STATISTICALLY

As mentioned in Chapter 19, despite the irregular spacing of the nontrivial Riemann zeta zeros, they do obey some laws. Rather like the primes, there's a formula involving a logarithm which tells us approximately how many zeros can be found between the horizontal axis and any given height. Unlike the primes, which tend to become increasingly sparse (according to a logarithmic law) the farther out we look along the number line, the zeta zeros tend to become increasingly *dense* (also according to a logarithmic law) the higher up we look in the critical strip. So if we think of the primes as thinning out in a way which can be loosely related to the uncoiling of a snailshell, we could similarly think of the zeta zeros as crowding together in a way which relates to the opposite – the winding in towards the centre.

With a simple transformation it's possible to "strip out" this logarithmic clustering effect. The critical line can be stretched in a particular logarithmic way [31] so that the zeros end up spaced on average a distance of 1 apart (but fluctuating greatly around this average).

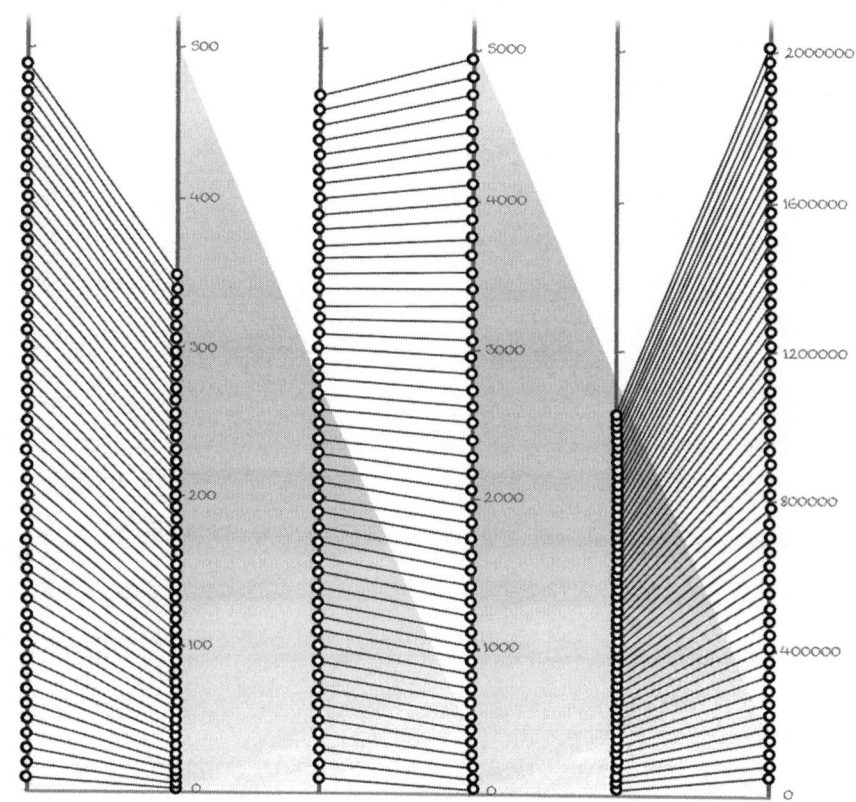

Here we see how this "stretching" works. Notice how up to about 3500, points gets pushed down the strip, but above that they get pushed up (at an ever increasing rate).

Having done this, we're in a better position to discuss the matter of randomness in the distribution of nontrivial zeros. It's in this "normalised" setting that random matrix theory is able to tell us something about the statistical "fingerprints" of the zeta zeros. If we look at the (new) spacings between pairs of zeros (particularly neighbouring pairs), collecting and statistically analysing this data, the same statistical patterns are found as those which occur in the spacings between eigenvalues in the spectra that tend to emerge from large GUE matrices (see Chapter 29).

One widely noted statistical tendency which the (normalized) zeta zeros and GUE eigenvalues have in common is a kind of "repulsion" which is evident in adjacent zeros or eigenvalues. It's as if they're trying to avoid each other! If we were looking at what are called *Poisson-distributed* random numbers (the Poisson distribution being one of the most commonly used probability distributions[32]) spaced out along a number line an average distance of 1 apart, there would be no correlation, the numbers all being entirely independent of each other, and we wouldn't see this "repulsive" phenomenon. As mentioned on page 32, Dyson's formula describing the distribution of gaps between eigenvalues in GUE spectra predicted this repulsion, as did Montgomery's "pair correlation" formula for the zeta zeros. So the "styles" of randomness inherent in the nontrivial zeta zeros and in the GUE spectral levels appear to have something in common. The more zeta zeros we can compute, the more precisely we can test this hypothesis. And all the computations performed to date have only strengthened the certainty of analytic number theorists and random matrix theorists that the GUE-zeta connection, formulated most generally as Montgomery's "GUE Hypothesis" is undeniable.

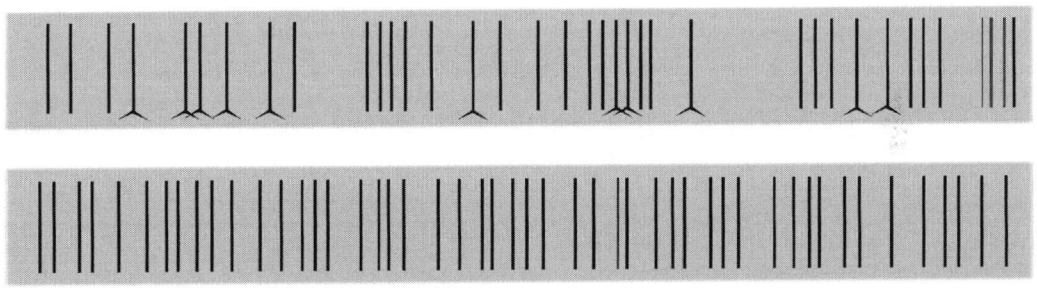

Upper spectrum: Poisson–distributed levels, no repulsion evident (the arrows indicate multiple levels too close to distinguish); Lower spectrum: GUE levels, or "normalised" zeta zeros, displaying repulsion.

Recall that from a quantum chaology perspective, this tendency for repulsion in the zeta "spectrum" indicates that the underlying classical dynamics associated with it should be chaotic. Poisson-distributed (uncorrelated) spectral levels would indicate underlying dynamics which are non-chaotic.

STOCHASTIC PHYSICS

Here's a word which is familiar to mathematicians and physicists, but to almost no one else: *stochastic*. This is sometimes used as a slightly more technical way of saying "random". There's more to it than that, but when physicists talk about "stochastic systems", this generally means physical systems involving randomness or random-like behaviour.

The fact that random matrix theory can be applied to the distribution of nontrivial Riemann zeta zeros hints that some kind of "stochasticity" is at work in the number system. I've mentioned a couple of times already that prime number theory has been linked to statistical mechanics (the oldest, most well-established branch of stochastic physics). Putting the details aside, the key point here is that the techniques of statistical physics, developed for studying unmanageably large aggregates of particles, have been found to be uncannily appropriate for studying the behaviour of the Riemann zeta function, the pure mathematical structure at the heart of the number system. Rather like the link with quantum chaology, this often produces a sense of surprise bordering on disorientation among those first exposed to it, as well as a kind of wonder or delight at this unexpected connection which seems to hint at a deeper, richer intertwining of physical and mathematical realities than anyone had previously imagined.

Other key areas of stochastic physics involve *Brownian motion, diffusion processes* and *entropy*. In each case, someone has found the physical theory curiously applicable to the distribution of prime numbers [33].

All of this, as well as calling into question the current doctrine concerning the relationship between the number system and physical reality, strengthens the curious sense of there being "something stochastic" underlying the (eternal, unchanging, carved-in-stone) number system.

This is perhaps best captured by that remark of Timothy Gowers which I quoted earlier:

> *"Although the prime numbers are rigidly determined, they somehow feel like experimental data."* [34]

Note the use of the word "feel". Mathematics in its formal, academic context requires the filtering out of anything connected with mathematicians' feelings about their subject matter. But occasionally (usually in articles, books and lectures aimed at a more popular audience), these feelings emerge. Here, Professor Gowers is telling us not what the distribution of primes *is*, but what it *feels like*. And although such an assertion is not intended to be rigorous (in the way a mathematical theorem or hypothesis should be), almost all number theorists would *know what he means*. We're grasping at something very subtle here.

chapter 34
number and time

"What then is time? If no one asks me, I know. If I wish to explain it to one that asketh, I know not." (Saint Augustine) [1]

Although time plays a major role in just about every branch of physics (in order for anything to *happen*, time must be involved), it has been surprisingly neglected as an object of study in its own right. Developments in 20th century physics such as relativity theory and quantum cosmology have given rise to some very interesting questions concerning the nature of time, but these have largely remained on the philosophical fringes of the subject. Philosophers have been debating the nature of time for centuries, but due to the highly specialised nature of the physics involved, today's philosophers are generally unable to fully engage with these new developments.

Recently, though, things have started to change. The branch of philosophy concerned with the nature of time has traditionally been called *temporality*. The word *temporology* has been introduced more recently in an attempt to capture the more interdisciplinary spirit of the current discussion about time which involves not just philosophers, but also physicists, psychologists and neuroscientists. The International Society for the Study of Time has been publishing a peer-reviewed journal and holding triennial conferences since 1966. Other temporology seminars have been held in Russia and Slovakia [2], and since 1984, Moscow State University has been hosting an "Institute for the Exploration of the Nature of Time" [3].

There have been similar developments in regard to consciousness. Philosophers have been discussing the nature and origins of consciousness for centuries, but scientists, having nothing to measure or quantify, have generally ignored this most fundamental (and, arguably, miraculous) of phenomena. However, an interdisciplinary field of "consciousness studies" involving philosophers, neuroscientists, physicists, psychologists, anthropologists, theologians, computational theorists and others has emerged in recent years, spawning a number of conferences[4] and at least one journal[5].

Time and consciousness are perhaps the most basic of mysteries, directly accessible to all of us. Yet, unless we're philosophers, we rarely give them any thought. Until recently, scientists have avoided these issues. When I was younger, I'd often wonder why no one was bothering to investigate them. Then, towards the very end of the 20th century, things began to happen.

Temporology and consciousness studies are both relatively new interdisciplinary fields, dealing with the most basic aspects of experience, with something which is "right under our noses" (although "behind our eyes" is perhaps a more apt description).

Part of the reason why there had been so little serious investigation into these matters (at least within Western science[6]), I think, is the difficulty in getting any kind of grip on them. The way our thoughts are structured is closely tied to how we relate to time, so trying to think about time or consciousness is inevitably going to present major conceptual difficulties. It's as if we're stuck inside something, trying to deduce what it looks like from the outside. Still, with a variety of perspectives and indirect methods, some steps forward have been made.

In both cases, we have an encouraging instance of the scientific establishment humbling itself, admitting that it doesn't have a clue what time or consciousness is, and proposing that everyone concerned, from whatever field, gets together and talks about it.

I feel that it's similarly high time for an interdisciplinary approach to number. This could involve mathematicians, philosophers concerned with the foundations of mathematics, physicists, psychologists, anthropologists and researchers from other fields who have something to add to the discussion of "what numbers really are" (or, more accurately, "what the number system really is"). We've reached a point now where we ought to admit that, as with time and consciousness, *we just don't know.*

As we'll see in the chapters to come, there seems to be something very deep and poorly understood going on which relates (1) human consciousness, (2) our perception of time, (3) the relationship between addition and multiplication, and (4) the mysterious nature of the prime numbers.

BEYOND THE TIMELINE

> *"I have sometimes thought that the profound mystery which envelops our conceptions relative to prime numbers depends upon the limitations of our faculties in regard to time, which like space may be in essence poly-dimensional and that this and other such sort of truths would become self-evident to a being whose mode of perception is according to superficially as opposed to our own limitation to linearly extended time."* (James Joseph Sylvester, 1888)[7]

Rewritten in contemporary English, Sylvester is saying something like this:

"I have sometimes thought that if we were able to perceive time in some multi-dimensional way, more like a surface than like a line, then the distribution of prime numbers and related mysteries would be entirely self-evident and would not seem at all mysterious to us."

James Sylvester was a competent poet as well as an accomplished mathematician. I feel that there's as much "poetic perception" in what he was trying to get across in this remark as there is mathematical insight. This is the earliest published statement I'm aware of which attempts to link the nature of time (or our perception of it) with the mysteries surrounding the distribution of prime numbers. Sylvester was speculating, of course, and we'll never know exactly what he meant, but we have a few possible clues.

Before proceeding, we should be clear about what is meant by "dimensions". People often get confused about this word. In the mathematical sciences, it has a precise meaning, but the word has also been adopted as a vague, descriptive term used in various cultural and psychological contexts – people talk about new dimensions of experience, of the mind, *etc.*, and a band might talk about adding a new dimension to their sound. There's also a strong science fiction connotation, bringing to mind fantastic images of future humans (or other entities) entering into or moving between "other dimensions".

So when people learn that Einstein put forward the idea of a four-dimensional reality, with time being the "fourth dimension", they can get extremely confused about what he meant by this. We're all capable of imagining a one-dimensional object (a line, or part of a line, straight or curvy) or a two-dimensional object (a plane, or part of a plane, flat or wavy), and we think of the space we move around in as three-dimensional. But it's very difficult to imagine how something can be "four-dimensional". People seem to imagine very different things (if they're able to imagine anything at all), often wrongly assuming that their particular mental image is what Einstein was talking about.

To make this as clear as possible, we'll first consider the fact that the word "dimension" gets used in another, particularly mundane, context. When people are talking about the size of a box or box-shaped object (a filing cabinet, suitcase or paving slab), they describe its "dimensions". Saying that something has "dimensions 12 feet × 20 feet × 5 feet" would describe this:

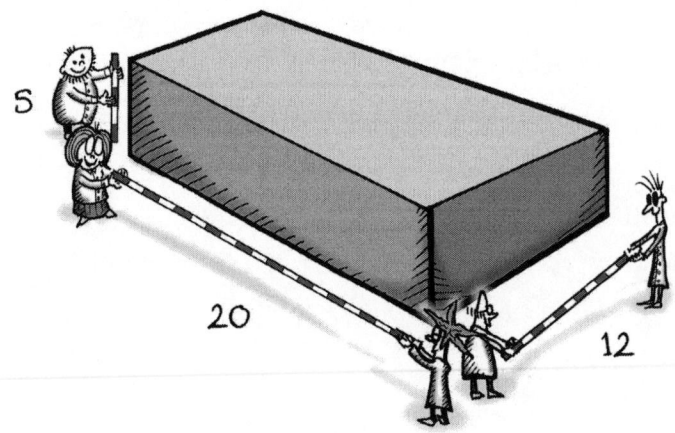

So this object has *three dimensions*. "Dimension" is really just another word for "measurement". We have three measurements. 12 feet is the measurement of how long one edge is, 20 feet is the measurement of another and 5 feet is the measurement of the remaining edge.

It's hard to imagine this thing having a *fourth* dimension. Notice the corners. At each, we have three right angles meeting. This is how you can define the dimension of a geometric space: *what's the largest number of lines that you can put through a point which are all at right angles to each other?* If you're looking at a line, there's only one line you can do this with (there can be no other lines at a right angle to it contained within the line itself, of course). If you're looking at a plane, there are two, but no more. To get a third, you have to leave the plane, in other words introduce "a third dimension" (a third direction that you can now take measurements in).

Because we treat time as if it were a line, Einstein was able to suggest that, in an abstract mathematical sense, this one-dimensional timeline could be joined to the three-dimensional space we appear to inhabit, in this way constructing a four-dimensional mathematical model which includes both time and space within a unified geometric framework. So *in this framework* time is indeed "the fourth dimension", but there are many possible mathematical constructions of four-, five-, six- and higher-dimensional spaces which have no particular relationship to time or physical space. They are purely abstract mathematical entities. These spaces can be thought of as consisting of points, and in each case the dimension tells us the number of "measurements" (numerical coordinates) we need to specify in order to describe the location of a point unambiguously.

A single measurement can locate any point in a one–dimensional space (line), whereas two are needed for a two-dimensional space (plane) and three are needed for a three–dimensional space.

In Einstein's four-dimensional "space-time" the points are called *events*. Each such "event" corresponds to a precise location in space at a precise moment. So if you need three measurements to describe a location in space, you'll need four (the original three, plus an exact time) to describe one of these points. That makes "space-time" a four-dimensional object.

The real number line is a one-dimensional number system. The complex plane is a two-dimensional number system. There are higher-dimensional number systems of various kinds which have been investigated in detail[8], but the most interesting objects of study seem to be found in the first two.

Now, back to Sylvester's remark. When he supposes time might be multi-dimensional (writing years before Einstein's work), this has nothing to do with "time as the fourth dimension". We in the West (Einstein included) have long been thinking about time as if it were a one-dimensional phenomenon, but Sylvester was suggesting that it might have another dimension (or more) – some kind of "sideways" time, perhaps?[9] This is extremely hard to think about in any coherent way as our thoughts are very much structured in relation to our cultural understanding of the passage of time.

There have been a number of scientists who have proposed speculative two-dimensional time models in order to explain certain problematic phenomena, although none of these have yet been assimilated by mainstream science[10].

At this point, it's worth briefly mentioning Donald Crowhurst, the ill-fated yachtsman who descended into madness in the Sargasso Sea while obsessing over the passage of time and one of Einstein's books on relativity theory. Before taking his ship's clock with him to the bottom of the sea, he wrote the following curious passages among the crazed philosophical ramblings with which he filled his log book during his last days:

"God's clock is not the same as our clock. He has an infinite amount of 'our' time. Ours has very nearly run out…"

"The Kingdom of God has an area measured not in square miles but in square hours. It is a kingdom with all the time in the world — we have used all the time available to us, and must now seek an imaginary sort of time."[11]

Sylvester's suggestion is relevant to a phenomenon we'll look at in more depth in Chapter 39 – all the "surprise" which we've encountered thus far. People are *surprised* by something when they weren't expecting it, or they were expecting something else. To me, this suggests that we're surprised by aspects of "our" number system because we've somehow been expecting the wrong thing – there's some fundamentally misguided perception involved. Sylvester seems to be hinting that the fundamental issue at stake here is our misperception of time (we're only seeing a line, but this line might be part of a wider "surface"). Because of this possible failing in our perception, he appears to be saying, the prime numbers look bafflingly disordered to us, but this is only because we're not seeing the bigger picture of which they're part.

Whether or not Sylvester was barking up the right tree, having read this far, I'd be surprised if you didn't share my feeling that there's *something* we're missing in relation to the primes and how they distribute among the counting numbers. If such a "something" were to become clear to us, it could result in a sudden leap forward in our understanding of the number system. And *time* does seem to be as good a starting point as any in seeking this missing piece of the puzzle.

PHYSICS CONNECTIONS

Time is something which physicists are more likely to be concerned with than are mathematicians. I've already made the simple observation that whenever anything *happens*, time must be involved. To put it very simply, physics generally deals with "things that happen", whereas mathematics is more concerned with "things that just *are*".

A lot of physics involves the development of mathematical models for various types of dynamical systems (which are "things that happen" in a particular sense). The equations involved very often contain a "t" variable, representing time as a numerical parameter. Such equations are intended to describe the "temporal evolution" of the system. Time clearly plays a major role in this kind of physics, but it's not the *subject* of it, more like the "track" along which it's occurring. Two issues in physics where time *is* the subject of investigation are the measurement of time and the "arrow of time".

The measurement of time turns out to be a extremely subtle matter, part of the wider field of *metrology*. We're now so used to devices which precisely keep time that this might seem surprising. The problem is, how do you test the accuracy of such a device? With another device, necessarily. And how do you test the accuracy of that? This way of thinking eventually leads us down into the quantum realm, where the measurement of time is defined in terms of (for example) the emissions from cæsium atoms. This is how physicists currently define a *second*: the amount of time it takes for 9 192 631 770 oscillations of a particular frequency of radiation associated with the spectrum of a cæsium atom to occur! In the end, we have to rely on the seeming regularity of the universe at that scale as we're basing our measurement of time on it. And that's just one aspect of what's involved.

Michel Planat is a metrologist working at the French National Centre for Scientific Research's FEMTO-ST institute. While part of the institute's Laboratory of the Physics and Metrology of Oscillators in the 2000s, his research concerning signal processing (very much applicable to practical realities like communication technology) led him, via such physical phenomena as *1/f noise* and *phase locking*, to such number theoretical matters as *Ramanujan sums*[12], *incomplete Gauss sums*[13], the *Möbius function*[14], the *von Mangoldt function*[15] and ultimately the Riemann Hypothesis. As a result of these connections, his work relates number theory to certain physics-related questions concerning the nature of time.

Some of the titles of his publications tell the story (I've put the number theory parts in italics and the physics parts in boldface):

"The impact of *prime number theory* on **frequency metrology**"

"**1/f noise, the measurement of time** and *number theory*"

"**1/f frequency noise in a communication receiver** and *the Riemann Hypothesis*"

"**Time measurements, 1/f noise of the oscillators** and *algebraic numbers*"

"*Ramanujan sums* for **signal processing of low frequency noise**"

"The *arithmetic* of **1/f noise in a phase locked loop**"

"*Arithmetical* **chaology** and the signatures of **1/f noise**"

"**Quantum 1/f noise in equilibrium**: from **Planck** to *Ramanujan*"

"On the *cyclotomic* **quantum** algebra of **time** perception"

The last of these papers is highly adventurous for the world of academic physics, venturing into speculation about the nature of time perception. Usually this would be considered part of psychology – notice that I haven't put "perception" in boldface as it's not traditionally something studied by physicists. Planat explains in the paper's introduction (and don't worry, you're not expected to understand this, but the language will give you a feeling for what's involved):

"*I develop the idea that time perception is the quantum counterpart to time measurement. Phase-locking and prime number theory were proposed as the unifying concepts for understanding the optimal synchronization of clocks and their 1/f frequency noise. Time perception is shown to depend on the thermodynamics of a quantum algebra of number and phase operators already proposed for quantum computational tasks, and to evolve according to a Hamiltonian mimicking Fechner's law. The mathematics is Bost and Connes' quantum model for prime numbers. The picture that emerges is a unique perception state above a critical temperature and plenty of them allowed below, which are parameterized by the symmetry group for the primitive roots of unity. Squeezing of phase fluctuations close to the phase transition temperature may play a role in memory encoding and conscious activity."*[16]

It's not at all clear what this is telling us about the nature of the number system, but it's yet another surprising example of physicists finding unexpected connections between their work and that of number theorists. It adds to the emerging sense that the number system *isn't what we thought it was* and could be taken as an indication that one way to further our understanding of number would be to reassess the way we conceptualise time.

The *arrow of time* is an issue which comes up in that hazy area where physics meets philosophy (hazy because philosophers are rarely in a position to learn any detailed physics and physicists usually too busy doing physics to be concerned with philosophising about it). We *experience* time as if it's flowing in a particular direction – "from past to future", not the other way. Time is conventionally modelled as a number line, so in what sense can we say that this line is directed from "left to right" or "right to left"?

Suppose you were a trans-cosmic entity with access to a kind of 3-d film of the entire history of the universe *and no other experience of it* (a supremely hypothetical situation, of course!). How would you know if you were playing the film forwards or backwards? Could you, by observing and attempting to deduce physical laws, determine which was the "correct" way to view it? Obviously, when *we* watch a film and see a gazelle running backwards or pieces of broken glass leaping from floor to table and assembling themselves into a bottle, we know that the film is running backwards. But that's based on our experience, the way our memory and consciousness operate and our intuitive grasp of physical laws. Without any of that, what could we deduce? Is there a "correct" direction for the arrow of time? As well as being relevant to the time-irreversible systems mentioned in Chapter 30, these questions are related to the issue of *causality*, the idea that the universe operates in terms of chains of "causes" and "effects". Are these things intrinsic to the nature of reality (are they "really there")? Or is this just the way we structure our perception?

The research of Moscow-based physicist Igor Volovich involves number theory in the form of *p-adic number systems*. This has led to the beginnings of an explanation of why we experience the world as if there were an "arrow of time"[17]. To get a flavour of this, we'll need to know what these "*p*-adic number systems" are about. It's worth looking at these in some detail, as they're involved in significant portion of the current literature which connects number theory and physics. "*p*-adic physics" is now sufficiently well established that there are regular international conferences being held on the subject[18].

p-adic numbers did get a brief mention in Chapter 16 when we were defining the real numbers. We'd constructed the system of rational numbers $\mathbb{Q}$ and then gone on to see that it's "full of holes", that is, it doesn't fill up the entire line – there are points on the number line that cannot be captured in the infinitely dense net that is $\mathbb{Q}$. These, I explained, correspond to the irrational numbers which, combined with the rationals, make up the system of real numbers $\mathbb{R}$.

I was intentionally vague about the definition of irrational numbers. This involves something called a *completion* which I described as a "*meaningful way to fill in the holes*". A completion requires a notion of *distance* between pairs of numbers in $\mathbb{Q}$. The usual completion, the one that leads to $\mathbb{R}$, involves the familiar notion of distance – you take the distance on the line between the points representing the two numbers, like measuring a distance with a ruler. This is the same as subtracting the smaller number from the larger:

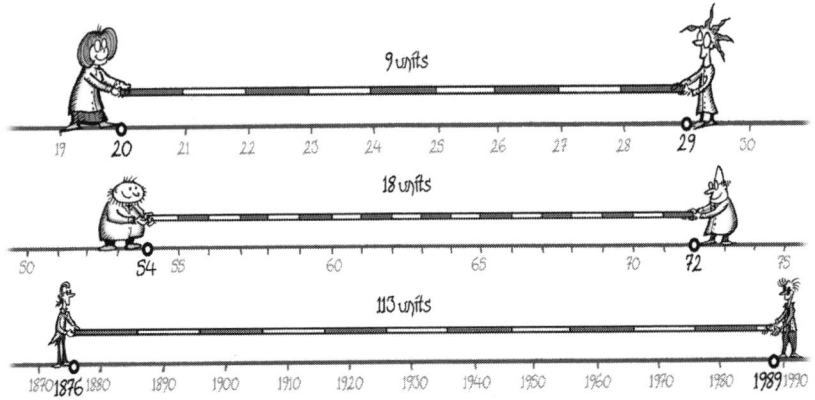

I mentioned that there are infinitely many other completions of Q, one for each prime number. Each leads to a different number system, these being known as the "2-adic", "3-adic", "5-adic", "7-adic", "11-adic",… number systems.

Each of these p-adic systems involves a completely different notion of distance between pairs of numbers in Q. This is "distance" in an abstract, generalised sense. It's not distance as you know it, but it still follows certain sensible rules such as (1) the distance from A to B is always the same as the distance from B to A, (2) the distance from a number to itself is always zero, and (3) distances are always either positive or zero, never negative.

Recall that rational numbers are basically the same thing as fractions, which means that we can write them as one counting number "over" another counting number, like this:

$$0.5 = \tfrac{1}{2} \qquad 0.75 = \tfrac{3}{4} \qquad 0.55555… = \tfrac{5}{9}$$

$$0.39 = \tfrac{39}{100} \qquad\qquad 0.427 = \tfrac{427}{1000}$$

Subtracting one fraction from another always produces a fraction. This means that the usual distance between a pair of rational numbers is always itself a rational number.

For our purposes, you won't really need to know how the "p-adic" notions of distance work, only that they involve:

(1) taking the usual distance between a pair of rational numbers,

(2) expressing this distance as a fraction – that is, one counting number over another,

(3) looking at how many times the prime number p shows up in the factorisations of these two counting numbers.

Here are three examples (fear not, you won't need to remember how this works):

$$2.0001 - 2 = 0.0001 = \frac{1}{10000} = \frac{1}{2^4 \times 5^4} \quad : 5\text{'s in denominator}$$

so the 5-adic distance between 2 and 2.0001 is 5^4, or <u>625</u>

$$783.75 - 118.25 = 665.5 = \frac{6655}{10} = \frac{11^3}{2} \quad : 11\text{'s in numerator}$$

so the 11-adic distance between 118.25 and 783.75 is $1/11^3$, or <u>0.000751...</u>

$$41.87 - 9.35 = 32.52 = \frac{3252}{100} = \frac{3 \times 271}{5^2} \quad : \text{no 2's involved}$$

so the 2-adic distance between 9.35 and 41.87 is 2^0, or <u>1</u>

Note that fractions are to be expressed in their simplest form, with no common factors occurring in the numerator and the denominator.

The number systems you get by "completing" $\mathbb{Q}$ according to these various number theoretic versions of "distance" – "2-adic", "3-adic", "5-adic", "7-adic", *etc.* – can seem utterly bizarre when you first encounter them. The whole notion of how you measure distance between points has been transformed beyond recognition (a bit like with the hyperbolic disc earlier, but more so). It's now all about the divisibility of numerators and denominators by a certain prime number, and the result is that points in $\mathbb{Q}$ which seem "close" in our usual number system can end up with huge distances between them in a *p*-adic number system, and points which seem a large "usual" distance apart can have a tiny "*p*-adic" distance between them. It's as if the number line has been thoroughly scrambled, with the prime number *p* being centrally involved in the specifics of this "scrambling".

At some point it the 20th century, it was observed that all physics then being developed was based on the usual concept of the number system – the real number line, or $\mathbb{R}$.

This was suddenly called into question – why did this particular completion of $\mathbb{Q}$ have a special status? Why use that one particular number system when there are infinitely many others to choose from? As a result of this kind of thinking, we now see the rapidly expanding field of "*p*-adic physics" where mathematical models developed to explain quantum mechanical phenomena, black hole cosmology, superstrings, dynamical systems, *etc.* use *p*-adic number systems (remember, the "*p*" here denotes some unspecified prime number).

But the question then arises as to *which* prime number we should be using. Why would "2-adic physics" be any more sensible or natural than, say, "137-adic physics"? One approach which has emerged is to combine *all* of these number systems – the "real" one $\mathbb{R}$, the 2-adic one, the 3-adic, 5-adic, 7-adic, 11-adic, ... – into a kind of "super-number system" called the adeles (briefly mentioned back on page 103). And so alongside *p*-adic physics we find an ever-expanding body of work on "adelic physics". Both tend to involve number theory far more directly than conventional physics does, simply because the primes are more immediately involved in the workings of the *p*-adic and adelic number systems than they are in $\mathbb{R}$.

Igor Volovich has published an extensive body of work on *p*-adic and adelic physics. As mentioned earlier, his has led him to suggest a context in which our experience of "the arrow of time" may have arisen[19]. This provides another link between number theory and fundamental issues concerning the nature of time.

Beyond this, a few theoretical physicists have been exploring the novel idea of *p*-adic (or adelic) *time*. After all, if time is going to be treated as a line, why should we use the real number system to measure it? We now have all these other possible ways we can understand the number line and the way it "interacts with itself" (based on the notion of distance being used). Because these systems work in terms of prime divisibility, number theoretical issues inevitably get involved in such conceptualisations of time. These ideas are being explored not as philosophical novelties, but as potential tools for physicists to use in their modelling of physical phenomena. And certain successes in theoretical physics provide evidence that *p*-adic and adelic time may turn out to be very useful models indeed[20].

chapter 35

addition and multiplication revisited

"The Riemann Hypothesis is the most basic connection between addition and multiplication that there is, so I think of it in the simplest terms as something really basic that we don't understand about the link between addition and multiplication." (J. Brian Conrey) [1]

"[The RH] is probably the most basic problem in mathematics, in the sense that it is the intertwining of addition and multiplication. It's a gaping hole in our understanding..." (Alain Connes) [2]

Conrey and Connes are suggesting that the mystery of the primes, epitomised by the difficulty of the RH, can ultimately be traced back to the uneasy relationship between addition and multiplication.

The simple act of counting can be related to time. To count five beans, you must go through a process something like this:

✩ I see a bean.

✩ I see another bean, which is not the same bean I just saw.

✩ I see yet another bean, which is not the same as either of the beans I've just seen.

✩ Again, I see another bean, and it's not the same as any of the others I've previously seen.

✩ I see a final bean, different from all the rest.

These five events happen in sequence, however rapidly. They occur in separate little segments of time. Those segments of time can all be put into a kind of correspondence with each other since *each of them includes an acknowledgement of seeing a (previously unacknowledged) bean.*

Arguably, if you arrange beans in the forms of a triangle, square, pentagon, *etc....*

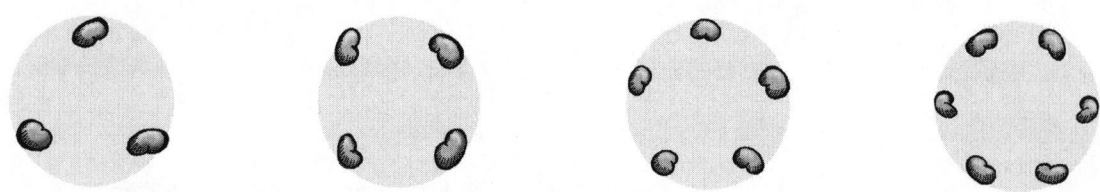

...then the brain can immediately identify the shape and thereby arrive at the correct "count" without having to do the counting. But that's not really counting, it's just recognising shapes and giving them numbers for names. Also, it only works for the first few counting numbers (and these are arguably the ones that are "hard-wired" into the brain, at least according to the research of Stanislas Dehaene[3]). Beyond this initial few you just have to count, and that necessarily involves the relevant number of little segments of time, all of which can be put in "conceptual correspondence" with each other.

We'll now consider addition in this light. If I say "five plus three", you say "eight", but you probably haven't actually *added* the numbers. You've remembered the answer to a fairly common question. It's similar to me asking "what do you get if you mix yellow and blue?" and you saying "green" – no calculation is involved. If I say "1909 plus 2317", then most people would need a pen, paper and a technique they've learned, but many would struggle if asked to explain clearly how/why this technique works. You might be aware of some mental arithmetic tricks that could be used, but these too would be found to be rather complicated if carefully examined.

To get back to basics and truly add 5 to 3 or 1909 to 2317, you'd need to count out five (or 1909) beans, or marks on paper, or whatever, then count out another three (or 2317), and finally count the whole lot to see how many there were. These acts of counting inevitably mean that each bean or mark receives a moment of acknowledgement, however brief that is. These moments of acknowledgement are necessarily separated in time, and they can be put into correspondence with each other as "similar events" (acknowledgements of a previously unacknowledged bean or mark).

As discussed at some length in Chapter 6, multiplication isn't as simple as addition. In fact it's a sort of "iteration" or compounding of additions. 4×7 is $7 + 7 + 7 + 7$ (more helpfully, $0 + 7 + 7 + 7 + 7$) so the 4 and 7 are playing different roles: the 7 is the thing being added, the 4 is the *number of times* it's being added (to zero). Despite the symmetry of "$4 \times 7 = 7 \times 4$" – meaning you can add four (to zero) seven times and get the same result – there's an asymmetry present in any *act* of multiplication, and it's somehow related to time. When we consider $4 + 7 = 7 + 4$, we find a more straightforward symmetry: 4 and 7 are both numbers being added to something. Their roles are identical on both sides of the equation.

> "*Multiplication is more complex* [than addition]. *When we multiply 2 × 3 we either take two threes and add them together, or we take three twos and add these together. In either case we make an operator out of one number and use this operator to reproduce copies of the other number.*" (Louis Kauffman)[4]

In *Uncle Petros and Goldbach's Conjecture*, author Apostolos Doxiadis makes the following point through his fictional mathematician character:

> "'*Multiplication is unnatural in the same sense as addition is natural. It is a contrived, second-order concept, no more really than a series of additions of equal elements. 3 × 5, for example, is nothing more than 5 + 5 + 5'… 'If multiplication is unnatural,' he continued, 'more so is the concept of "prime number" that springs directly from it. The extreme difficulty of the basic problems related to the primes is in fact a direct outcome of this. The reason there is no visible pattern in their distribution is that the very notion of multiplication – and thus of primes – is unnecessarily complex. This is the basic premise…*'"[5]

English-speaking people still generally prefer to say "seven times twelve" (rather than "seven multiplied by twelve"). This makes sense if we write

$$84 = 7 \times 12 = 0 + 12 + 12 + 12 + 12 + 12 + 12 + 12$$

that is, 12 added to zero 7 times. In a slightly outdated use of language, this is "seven times twelve"– consider the statement "seven times she rang the bell" and then think "seven times twelve got added (to zero)".

Similarly,

$$84 = 12 \times 7 = 0 + 7 + 7 + 7 + 7 + 7 + 7 + 7 + 7 + 7 + 7 + 7 + 7$$

or "twelve times seven". As observed in Chapter 6, generations of British school children have informally called the multiplication tables they're encouraged to memorise "times tables".

Any experience of "three times seven" will involve either three instances of something with seven elements or seven instances of something with three elements. Let's just stick with the first option for now. As far as I'm able to see, the three instances of the seven elements will always involve patterns of events separated in time, the three of which can be "put in correspondence" or seen as in some way "similar". For example, you could lay out three rows of seven beans. Or you could just count out seven beans into a pile three times.

Or you could take seven steps in a certain direction, and then again, and then again.

...or make seven marks on something three times:

With each of these approaches, there is a pattern of events being repeated three times, hence occurring in three separate chunks of time (laying out a row of seven beans, counting out seven beans into a pile, taking seven steps, making seven marks).

There are situations which are less obvious. Suppose that you're presented with three piles of seven beans. That's a representation of 3×7 which appears not to involve three "corresponding patterns of events separated in time". But, experientially, for this arrangement of beans to be "3×7" *to you*, you must have counted the three piles. However quickly you did this, your brain was effectively working as follows:

"I see a pile of beans. I see another pile of beans distinct from the pile I just saw. I see yet another pile of beans distinct from both the previous piles I've seen. I see that there are no other piles present. I'll now check the three piles (I've counted them and found that there are three) to see that each contains seven beans."

The act of counting piles means that each pile must be acknowledged as a thing in itself – distinct from each of the other piles. Each of these three acknowledgements must be separated in time:

And each acknowledgement itself involves seven elements (beans in the pile).

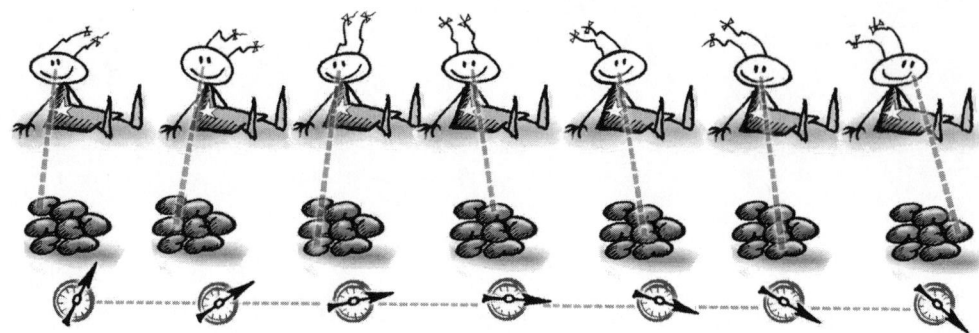

Suppose I lay out five piles of nine beans. Then, looking back at the time continuum, we can identify five separate intervals, each sharing the quality that *it contains a representation of me laying out a pile of nine beans* (imagine we'd filmed me laying out the piles and were now cutting the film up with scissors). But to confirm that these five "chunks of time" are truly in correspondence, we must examine the contents of each, checking that it indeed involves me making a pile of nine beans. To do *that*, we're necessarily counting (even if we've had to imagine ourselves temporarily stepping outside the time continuum in order to do this). That counting (which involves the identification of a distinct bean nine times) can also be understood in terms of the cutting up of time (in this case, separating out nine little segments).

If you think about addition in this time-based way, you'll see that it involves little more than just counting. Multiplication, though, is more like "counting-within-counting" (or "counting × counting" [6]).

The interesting thing here is that we're visualising it like this: You're filming me laying out the five piles of nine beans. Then you're suddenly in the "editing suite" cutting out five strips of film (which represent chunks of time), each showing me laying out a pile of nine beans. If the film represents ordinary time, then you've stepped into a "hyper-contextual" time continuum (rather like those novel cartoon sequences where animated characters escape from their cartoon and begin interacting with their animators[7]).

Each of these five strips of film must now be examined to confirm that it indeed shows me laying out of a pile of nine beans. But that involves you in the editing suite, counting, and hence, for each strip, the occurrence of nine "similar events" (you identifying a distinct bean being laid out on the film by me).

To be really thorough, then, someone would have to film *you* in the editing suite, five times, each time counting the beans on one of the strips of film. Finally, stepping out of the editing suite into the next level of "hyper-contextual time", you could cut up each of the five new films into nine pieces, each showing you looking at a piece of the original film, identifying me laying out a distinct bean. So, to really make sense of "five times nine" beans in terms of time, we end up with films within films, showing each of the 45 beans being individually identified, grouped into five lots of nine.

So, looking at counting and addition in this way, you're one stage removed from the actual beans, steps or marks: you're in the "editing suite", examining what appear to be corresponding events in the time continuum we associate with "reality". With multiplication, you're *two* stages removed: you're in an editing suite looking at film of yourself in another editing suite looking at film of "reality"!

All of this suggests to me that numbers and counting (before multiplication is introduced) are already strongly linked to our experience of time. But when you start multiplying (which involves counting not the usual objects of experience, but *numbers of those objects*), the involvement of time gets somehow compounded on itself. It's hard to think or talk about this coherently as our thought and language structures are shaped by our experience of the passage of time, but I suspect that J.J. Sylvester might have been thinking along these lines when he suggested that the mystery of the primes might be self-evident if we were able to experience a more surface-like (rather than linear) time.

The introduction of multiplication, then, seems to be an important conceptual threshold where the relationship between time and number fundamentally changes. Also, when we introduce multiplication, we immediately have the concept of *divisibility* (35 is divisible by 7 because $5 \times 7 = 35$ and 21 is divisible by 7 because $3 \times 7 = 21$) and, with it, the possibility of defining prime numbers. This is the point where we can think of the primes as suddenly collapsing down from the "free-floating clusters" scenario (as pictured in Chapter 6) into the rigid grid of a number system. The positions on the number line where they end up have an absolute inevitability about them and yet, when looked at on a large enough scale, they produce a mysteriously irregular "splatter". After many centuries of familiarity with the primes, humans came to know (via Bernhard Riemann's work) that this splatter has a beautiful, unexpected harmonic understructure. Just over a century later, the link with quantum chaology was uncovered and the possibility of an underlying "Riemann dynamics" was proposed. But none of this would be possible without the Rubicon of multiplication being crossed, *allowing the objects of your counting to be numbers themselves*.

It could be that I'm oversimplifying matters with this analysis, as it's possible to demonstrate multiplication by means of a wholly geometric procedure (hence just space is involved, rather than time). This challenges my conclusion that it necessarily involves "counting-within-counting" or "time compounded on itself". But even then, time is arguably implicit – Appendix 15 will attempt to explain, if you're interested.

Also, there's something which I may have glossed over in the "editing suite" visualisation. If we'd used rows of beans placed next to each other rather than piles, then it would arguably be possible just to count a single row, instantly recognising that the others contained the same number of beans. But the recognition of this correspondence between rows involves subtle assumptions about the nature of space, time and perception.

Finally, and importantly, a lot of what I've written rests on the potentially problematic notion of "sameness", "similarity" or "correspondence" of events, with the category "events" being so broad as to be almost indefinable. Clearly, this notion is very much tied up with natural language and the way we use it to separate our experience of the world into nameable categories (what I would call "discrete countable objects"). This "chunking" of the world is, I believe, the other side of the coin to the act of counting (the two processes are "mutually arising", in the language of Buddhist philosophy) – this will be explored further in Chapter 37.

THE WIDER ARGUMENT

Our inability to completely understand the distribution of prime numbers is evident in our inability to prove the Riemann Hypothesis, the latter being described at the beginning of this chapter by two eminent mathematicians as *an inability to properly understand the relationship between addition and multiplication*. And as this relationship appears to have something to do with time, it might be the case that we'll remain unable to prove the RH (or really understand the number system) until we've reassessed our way of conceptualising time. Sylvester's fascinating suggestion is certainly in keeping with this possibility.

Here's a more complete version of what Connes was quoted as saying:

> "[The Riemann Hypothesis] *is probably the most basic problem in mathematics, in the sense that it is the intertwining of addition and multiplication. It's a gaping hole in our understanding, because until we really understand it we cannot say that we understand the line. Even the line is still extraordinarily mysterious.*"[8]

Western civilisation, unique in its development of the study of number to the point where the Riemann zeta function could be described and the Riemann Hypothesis hypothesised, is also unique in *its characterisation of time as a line* – a featureless, homogeneous continuum. Prior to Western contact, other cultures have tended to conceptualise (and seemingly experience) time as something more multifaceted, fluid or cyclic. Western science's striving to measure and quantify time (along with everything else) has led to a conceptual link being made between time and the real number line. Relatively recently, though, the mathematical community has become acquainted with the *p*-adic number systems which, in a sense, inhabit the same line as the reals, but which mathematically twist it up in mind-boggling ways (at least to *our* minds). I believe it's this *p*-adic (and adelic) mathematics which Connes is alluding to when he says that we can't claim to understand the line.

If we remain convinced that time has a line-like structure, as children in school history lessons have been led to believe for many years (everyone's familiar with the "timeline"), then it follows that we cannot claim to understand the nature of time. Schoolchildren encounter a similar line in their maths lessons as the "number line". The relatively tiny proportion who end up studying higher mathematics may learn about the conceptual foundations behind this number line, that is, about the "axiomatic system" underlying the field of real numbers $\mathbb{R}$[9], but despite the informed opinion of Professor Connes, they're unlikely to be shown anything about the number line in their studies to suggest that it could be "extraordinarily mysterious". Likewise those who go on to study physics would be unlikely to become aware of the "extraordinarily mysterious" nature of the time continuum which the "*t*" variable peppering their calculations refers to.

The rise of p-adic and adelic physics could eventually change this situation, though, and (in the long term) might even lead to a change in the way we commonly relate to time.

Chapter 36

number and time revisited

The appearance of some exotic mathematical structures called *type III factors* in Alain Connes' groundbreaking number theory work[1] suggests an interesting possibility for linking time- and number-related concepts. As it would take another book this size to explain them properly, though, I'll have to limit myself to impressionistic descriptions.

Type III factors (which have nothing to do with *prime* factors, I should clarify) are examples of what are known as *noncommutative algebras*. Some readers will be unfamiliar with the use of the plural "algebras". Most people think of *algebra* as that part of mathematics involving *x*'s, *y*'s and such things (as presented to them in school). In fact, while mathematicians consider this "algebra" to be a branch of their discipline, it's just one example of *an* algebra. Other examples are matrix algebra, *Boolean algebra*, *Clifford algebras*, *Temperley–Lieb algebras*, *Hopf algebras* and *von Neumann algebras* (there are *many* more types). Some of these are *commutative* algebras. That means, roughly, that when you multiply two things together, the order in which you multiply doesn't matter (all algebras involve something resembling multiplication). That's what we're accustomed to with our usual multiplication of numbers: $4 \times 19 = 19 \times 4$. The type of algebra taught to teenagers in schools is also commutative. But some algebras are *noncommutative*, meaning that you can get different outcomes if you reverse the order of multiplication.

You might recall from the discussion of matrices in Chapter 29 that matrix multiplication is noncommutative. Matrix algebra is the first example of a noncommutative algebra that you'd normally meet in a mathematics education. Type III factors are much more recent, mathematically specialised and difficult to understand.

What makes type III factors so interesting is that unlike any other algebras which had previously been studied, they come with a *built-in dynamical quality*. As Connes has written, such algebras "*...evolve with time!*"[2]. By this, he certainly didn't mean "with time" in the sense of familiar historical or clock time. He wasn't suggesting that if you study these algebras, you'll find that they can change from one day to the next! Rather, the algebras have a built-in structure which uncannily resembles a mathematical model of a dynamical system. A dynamical system, recall, is a structure which evolves with time – a mechanical clock, for example, some billiard balls bouncing around a table, a pendulum or the solar system. It's admittedly hard to grasp this idea of "an algebra with a built-in dynamical quality" if you're not well versed in higher mathematics (it's hard enough if you *are*). Mathematicians would describe these algebras as having a *canonical flow*. A *flow* is a particular type of dynamical system. Think of those weather charts covered with little arrows indicating direction and strength of airflow, or navigational charts similarly displaying oceanic currents – these at least partially convey the notion of a flow.

What kind of time this involves is not at all clear. It's certainly an abstract kind, as opposed to one which is directly experienced. It could perhaps be called "imaginary" time (but in the colloquial sense rather than in the sense of complex numbers).

The best way I've found to convey this concept of "flowing algebras" is to invoke the notion of the "mathematical landscape". Mathematicians often describe their subject in terms of a shared conceptual "landscape" which they're collectively exploring and mapping in minute detail. Some areas are already very well mapped, others have yet to be discovered. Until recently, it was as if the landscape had been completely static – full of endlessly fascinating and intricate topographic, "geological" and other such features, but *nothing that moves*. The "mathematical landscape" had been thought of as a timeless, unchanging environment. When Alain Connes discovered this property of type III factors in the 1970s, it was almost like the first time someone had seen something moving in the mathematical landscape – as if an explorer had come across rivers, waterfalls and whirlpools in a land previously believed to be motionless.

Once more, I must stress that this "motion" doesn't occur in the same time continuum as the one in which you're reading this book. Think of watching a film. While the film is running, the characters (who may have never existed in any "real life" form) are seen to be living through and interacting in a sort of "fictional time". It's separate from "our" time even though it resembles it in many ways. If you leave the film and come back a few minutes later, the characters might be several years in their future (or past). We're generally very comfortable with this kind of fictional time. So it might help to think of the time in which these algebras called "type III factors" flow as a kind of fictional time. The fascinating difference is that unlike a film or book, *no one created this*. It's just *there*, somehow built into the structure of reality. And importantly, Connes' work[3] (and a few other people's since) has related these curious "flowing algebras" to the Riemann Hypothesis.

Physicists have taken familiar dynamical systems like spinning tops and abstracted them into mathematical models which can be described with a few lines of symbols. For these models to be meaningful, there must be some shared context we can agree on in which to interpret these symbols. In the case of a spinning top this doesn't really present problems because we have a physical system (an actual spinning top in this case) which we can observe, measure and otherwise study.

The behaviour of this actual physical system should correspond to the mathematical symbols if they're appropriately interpreted. So, for this example, we have a realisation of a mathematical model which we can see and touch. Some of the symbols used will correspond to distances in space, angles and periods of time which are physically measurable. We can carry out the measurements with our rulers, protractors and stopwatches and everything should correspond nicely between the mathematics of the model and the measurements taken from the physical realisation. There's been much philosophical discussion as to why mathematics is able to correspond so closely to a physical system in this way, but that's not what concerns us here.

Things become interesting when dynamical systems theory is extended beyond the realms of physical possibility. It's perfectly feasible to construct a mathematical description of "something happening" in the same format as mathematical models of physically possible dynamical systems (like spinning tops) but where *no physical representation of this is possible*.

To get some sense of this, we'll start by looking at static (as opposed to dynamic) mathematical objects. I mentioned back in Chapter 27 that mathematicians can discuss a "17-dimensional hypersphere" (that's just an arbitrary example) with a complete agreement as to what they mean. They've taken the mathematical description of a three-dimensional sphere and extended it into physically unrepresentable dimensions.

So if someone were to ask "where is this 17-dimensional hypersphere?", we would have to answer "in the shared imagination of certain humans". A Winnie-the-Pooh also exists in the shared imagination of certain humans. The difference between a 17-dimensional hypersphere and Winnie-the-Pooh, though, is that we can all agree exactly on every imaginable property of the hypersphere (via mathematical symbols to be interpreted in some agreed, shared context). If we very loosely describe the human imagination as a kind of "mental hyperspace", then the 17-dimensional hypersphere could be said to live in this mental hyperspace.

Where would you find an ordinary sphere? In the physical world you'll find numerous imperfect representations of the "ideal mathematical sphere": a tennis ball, a soap bubble, the Moon, a ball bearing. These are all slightly different "spheres". Meanwhile, an ideal mathematical sphere exists in the mental hyperspace of every mathematician and it's always *exactly the same sphere*.

Now, where would you find a 17-dimensional hypersphere? Again, in various mathematicians' mental hyperspaces. And again, these will all be exactly the same 17-dimensional hypersphere. Everyone concerned can agree on all of its properties. But unlike the 3-d case, you can't make slightly imperfect physical seventeen-dimensional hyperspheres out of nylon, soap, rock or metal.

In other words, the possibility of physical representation is gone. The only context in which this thing can exist is the mathematicians' mental hyperspace. The outward symbols describing this mathematical "object" can be written in chalk, ink or pixels. We can change the font, colour, size or even the choice of symbols – none of this matters. They have no intrinsic meaning or content. They only become meaningful in the "mathematical imagination", that shared region of all our mental hyperspaces which requires mathematical training to gain access to. So, if the mathematical entity which we describe as a "17-dimensional hypersphere" could be said to exist, then it exists in the mathematical imagination – a specialised type of "collective imagination". The fact that mathematics is the same for everyone makes me think of the situation like this:

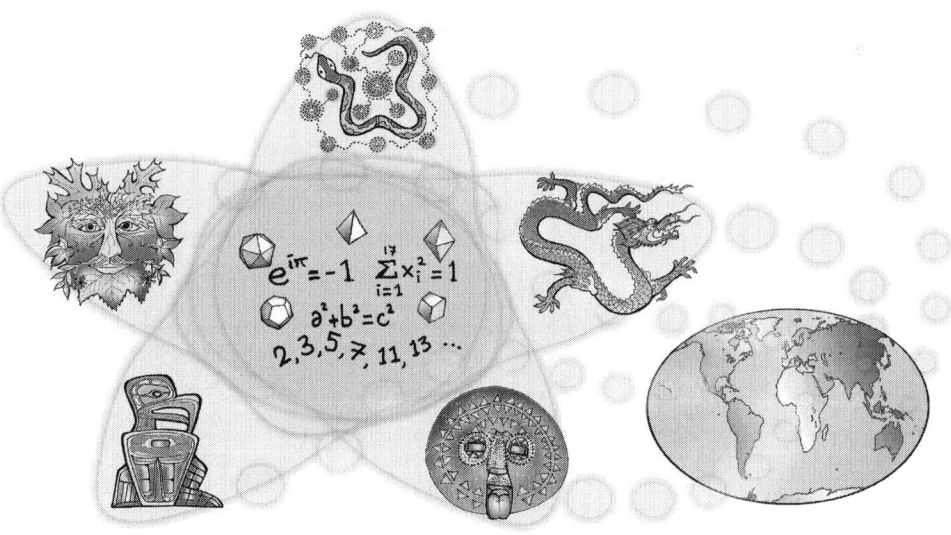

We'll now shift our attention from static geometric objects back to dynamical systems. Many commonly studied dynamical systems can be modelled by a *Hamiltonian flow*. Examples of these can be found in the laboratory (an oscillating spring, a pendulum mechanism). A spinning top, a ball rolling down a hill, a moon orbiting a planet – these too would be examples. You find these things in physical space-time. Despite certain complications like friction and limitations in the accuracy of our measurements, such "systems" can be measured with rulers, protractors and clocks and the measurements will accord with the mathematical model. Any such physical system corresponds to an ideal mathematical Hamiltonian flow which could potentially be present in numerous physicists' imaginations. If a number of these people individually attempted to construct a physical example based on the mathematical model, they would all necessarily produce imperfect representations[4]. But the mathematical model would (in some sense) be *exactly the same* in all of their imaginations.

Note that unlike the ordinary sphere or the 17-dimensional hypersphere, there's a time dimension involved in a Hamiltonian flow. This isn't a problem because we have a physical system, visible in physical space and evolving in physically experienced time. We can construct a model in our minds (mathematics) or possibly in the laboratory (engineering). In our shared "mental hyperspace", the model evolves *precisely* according to the equations. In the lab, the system being modelled will evolve *approximately* according to the equations, within a certain range of accuracy we find acceptable. We can measure with our rulers, protractors and clocks, and our measurements should accord with the mathematical model.

Now, where would you find a type III factor? It seems certain that we can't make a physical model of such a thing. The obvious answer, then (as for the hypersphere), would be "in certain mathematicians' mental hyperspace". It's the same type III factor for everyone, as if there's just one existing in some shared region of mind, rather than different copies existing in numerous minds.

The difference between a type III factor and a 17-dimensional hypersphere, though, is that the type III factor "evolves with time". There's a time dimension involved somehow, a kind of "fictional" time dimension. So the type III factor should really be said to exist in the mathematicians' (shared) *mental hyper-space-time*.

Like a Hamiltonian flow, the type III factor is in some sense *flowing*. The mathematics describing this flow is not that far removed from the mathematics of physically realisable flows. Yet (at least as far as we know), no physical model can be manufactured to "represent" a type III factor. Assuming no such thing is physically possible, we have here an example of something "flowing" in the mathematician's imagination, but which cannot flow in physical reality.

Considering the role which type III factors play in Connes' promising approach to the Riemann Hypothesis (as well as in Michel Lapidus' recent research – see Chapter 25) and then going back to James Sylvester's fascinating suggestion...

> "*I have sometimes thought that the profound mystery which envelops our conceptions relative to prime numbers depends upon the limitations of our faculties in regard to time, which like space may be in essence poly-dimensional and that this and other such sort of truths would become self-evident to a being whose mode of perception is according to superficially as opposed to our own limitation to linearly extended time.*" [5]

...it seems plausible that humanity will not be able to clear up the various confounding issues associated with the primes (currently embodied by the RH) until our understanding of time has been appropriately expanded or adjusted.

The sages and priests of many ancient civilisations appear to have been concerned with time. The development of calendars for practical, agricultural purposes is understandable, but if we look at, say, the complex set of interlocking Mayan calendric systems or the philosophy underlying the ancient Chinese *I Ching*, there appears to be a much deeper, less mundane concern – the nature, the "quality" or even the "structure" of time, as opposed to just its measurement. Recall that Carl Gauss described mathematics as "the Queen of the Sciences" and number theory as the "Queen of Mathematics". Could it be that number theorists (arguably the inner priesthood of today's "scientistic" [6] civilisation) are unintentionally laying the groundwork for a similarly motivated project – the exploration of the ultimate nature, structure and "meaning" of time? Perhaps in the distant future, if there are still humans around to think about such things, they'll realise that all of humanity's probing into the inner workings of the number system was just an unintended exploration of the "deep structure of time". But of course this is just wild speculation.

MORE DYNAMICAL SYSTEMS

I explained in Chapter 30 that, in connection with their quantum chaology work, Michael Berry and Jonathan Keating have proposed the existence of a dynamical system which could account for the spectral nature of the nontrivial Riemann zeta zeros, the so-called *Riemann dynamics* (as yet undiscovered).

Also, as we've just seen, as well as being noncommutative algebras, the type III factors studied by Alain Connes can be seen as dynamical systems (despite not being physically realisable) which now appear to be deeply linked to the structure of the number system.

There have been at least two other approaches to the Riemann Hypothesis which involve dynamical systems. Christopher Deninger has approached the problem using flows on *foliated spaces* [7]. I mentioned in Chapter 25 that Michel Lapidus (whose work overlaps in certain ways with elements of Berry and Keating's, Connes' and Deninger's) has proposed a "noncommutative flow on a moduli space of fractal membranes" in order to shed new light on the mysterious nature of the Riemann zeros[8].

Recall that dynamical systems involve (loosely speaking) "things happening with the passage of time", whether it's "physical" time or some kind of abstract/mathematical/fictional time. The proliferation of approaches to the RH involving dynamical systems (the interrelations between which are still not entirely clear) arguably reinforces the possibility that we won't fully understand the distribution of prime numbers until our understanding of time has been appropriately extended or amended.

Humanity's thinking about both time and number has evolved considerably over the centuries, and we have every reason to believe that this will continue, perhaps even accelerate. Everything discussed in the last few chapters makes me suspect that the next big shifts in our understandings of time and of number are likely to occur together.

LI, TIME AND NUMBER

I sometimes find myself comparing the sequence of prime numbers to an old, gnarled oak tree. It's characterised by its irregularity. It looks "wrong" to the kind of mind that wants everything arranged in neatly aligned stacks and geometric patterns. I know this because (left unchecked) I have that kind of mind. And yet even I can see that it has a perfection, beauty and grandeur.

In his last book, *Tao: The Watercourse Way*, Alan Watts explains that in traditional Chinese calligraphy, hesitating, hurrying or attempting to correct what has been written will inevitably lead to inferior work[9]. Calligraphy done well should almost take on a life of its own "*as a river, by following the lines of least resistance, makes elegant curves*". He compares the resulting calligraphic beauty to that which we see in flowing water, clouds, flames and "*weavings of smoke in sunlight*". To the ancient Chinese, this type of beauty was known as *li*. The ideographic symbol for *li* was originally associated with wood grain and patterns in jade, although it later took on a wider meaning as a kind of principle or meaning behind things (Joseph Needham, the influential scholar of ancient Chinese culture, translated it simply as "organic pattern"). Along similar lines, Watts points out that the Chinese notion of elegance is captured by *feng-liu,* which literally means "the flowing of wind". The forms of smoke, mist and clouds are able to provide a partial visual representation of the otherwise invisible *li* of air flows.

Later in the same book, Watts returns to the concept of *li,* describing it as a type of order which is easily recognisable but impossible to define because it involves far too many variables. He contrasts this "organic order" with more rigid legal and mechanical types of order, and gives further examples including the forms of trees, frost patterns on a window pane or scatterings of pebbles on a seashore. Europeans, Watts observes, took much longer than the Chinese to acknowledge and appreciate this type of beauty, noting that landscape painting was being practiced in China for centuries before Europeans could see any point in it. These days, photographers and painters, as well as abstract artists, draw on these organic types of form, the appeal of which is seemingly universal, while aesthetic theorists and art critics struggle to provide a clear account of why this is. Any attempt to reduce the sense of beauty to a well-defined geometric pattern misses the point: "*Geometrization always reduces natural form to something less than itself, to an oversimplification and rigidity which screens out the dancing curvaceousness of nature.*"

The various behaviours of air and water, frost, clouds, smoke, vapours, *etc.* are all governed by relatively simple physical laws, yet they often generate incredibly complex patterns. These can be partially (but never completely) modelled by means of the mathematics associated with chaos theory, fractal geometries and "self-similarity". Incidentally, these concepts of Western origin only arose towards the end of the 20th century and were not at all widely known in the mid-70s when Alan Watts wrote his book. Millennia earlier, the ancient Chinese mind was able to recognise the beautiful symmetry which we now call "self-similarity" and which we can find manifestations of in clouds, mountains and cauliflowers, among many other things. While the Chinese gave a name to this particular kind of beauty and cultivated it in their artwork, Western

civilisation had to wait for nonlinear mathematics and science to evolve to a point where it could be at least partially quantified.

There have been attempts by modern academics to analyse visual beauty, the aim being to discover what the mind is responding to when it finds something beautiful. Such efforts often involve some quantitative analysis of the subject matter in terms of proportion and symmetry. The underlying hypothesis is, very roughly, that the brain or mind has some kind of built-in "beauty receptors" which work (at least partly) mathematically. Although I feel that this approach is largely missing the point, we should consider the possibility that any such "beauty receptors" could be the products of enculturation as much as of "hard-wired" brain function. After all, notions of beauty can vary greatly between cultures. If they *are* predominantly culture-dependent (rather than hard-wired), then the ancient Chinese appear to have cultivated a very different type of "beauty receptors", as evidenced by their intuitive grasp of the fractal and self-similar geometries which the West, until relatively recently, almost entirely overlooked. This might begin to explain the differing natures of traditional Chinese and European aesthetics.

Whether or not there's any basis in reality for such a vague cultural/psychological/ aesthetic assertion, I feel that, at least in some metaphorical sense, the type of beauty which some people are able to find in the distribution of primes appeals to the ancient-Chinese-style "beauty receptors" in the mind. From this perspective, the expressions of alarm and "distaste" at the wild, chaotic, unpredictable nature of the primes would be the result of the dominant Western-style "receptors" being unable to attune to this kind of beauty.

Here are some examples of *li* taken from David Wade's superb little book on the subject[10]:

upper-left: chestnut bark; upper-right: a neuron; lower-left: fluid convection; lower-right: red cabbage

And here are some images generated from the distribution of primes or the Riemann zeta function which display *li*-like qualities:

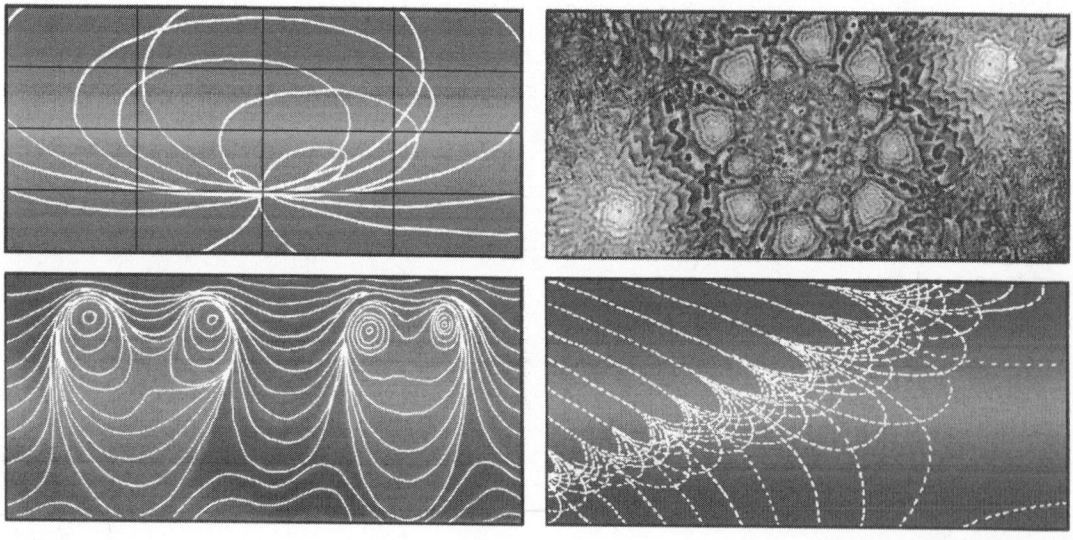

But what has this got to do with *time*?

In David Wade's book, he mentions that many examples of *li* can be characterised as *"frozen imprints of causative processes"*. That is, we're looking at something which was at one time happening and then somehow got "frozen in time". With geological *li*, the things we're looking at are still in motion (formation of mountains, movements of tectonic plates, *etc.*), but the time scale is so huge relative to human experience that they *appear* frozen to us. In general, with the types of *li* Wade is referring to, there's been activity which has (or appears to have) stopped, but some "imprint" of sorts has been left as evidence of this activity.

If the distribution of primes displays *li*, then this raises the (admittedly bizarre) question of whether this is evidence of some "causative process" underlying the number system — something which "happened" and then froze in that very particular shape. But any such "something" couldn't have happened in the past tense of historical/clock/"real" time. This would have to involve some other kind of abstract/mathematical/"fictional" time. Could this possibility be related to the curious applicability of dynamical systems to the Riemann Hypothesis? Once again, I'm drifting into the realms of wild speculation. Still, it's an attractively peculiar notion, almost suggesting something like a "creation myth" for the number system, an explanation of "how it got like that" in terms of a "fictional" process.

Interestingly, in another book, Alan Watts wrote...

> "*...for movement and rhythm are of the essence of all things lovable. In sculpture, architecture, and painting the finished form stands still, but even so the eye finds pleasure in the form only when in contains a certain lack of symmetry, when, frozen in stone as it may be, it looks as if it were in the midst of motion.*" [11]

When discussing this subject matter with non-mathematicians, I'm occasionally asked about the possible roles of chaos theory and/or fractal geometry in the distribution of primes. We've already seen a link with quantum chaos, although the connection with the actual distribution of primes (rather than the spacing of nontrivial Riemann zeta zeros) is somewhat indirect. Links with fractal geometry have also appeared in recent years[12] although they're rather obscure and not currently seen as being of any great mathematical importance. Michel Lapidus' work (where the distribution of primes is related to a "noncommutative flow" on a "moduli space of fractal membranes") is perhaps the most substantial effort in this direction as of 2013.

TIME, MUSIC AND PRIMES

Of all the arts, music has the closest relationship with time. I've heard geometry described as being the study of "the unfolding of number in space" and music as "the unfolding of number in time". Music theory and the physics of sound tell us that both harmony and rhythm can be described in terms of the subdivision of time intervals into particular numbers of equal parts. This leads us back to the belief system of the cult of Pythagoras (founded in the 6th century BCE by arguably the first mathematician) which had as one of its central tenets the mystical claim that "all is number". The Pythagoreans were particularly interested in musical harmony as an expression of the divine properties of the interrelationships between pairs of counting numbers.

It's perhaps not surprising that prime numbers show up in music theory. (We've already come across musical metaphors arising in connection with the zeta zeros, spiral waves, *etc.*, but I'm now referring to actual, rather than metaphorical, music.) The primes' role in music theory is not a large or obvious one. It's really only when you get into *tuning theory*, in particular the theory of *just intonation* – the sort of (surprisingly complicated) thing which esteemed tuners of concert pianos theorise about – that they start to show up.

At a less fundamental level, a few composers (Olivier Messiaen perhaps the best known) have attempted to use the irregularities of the sequence of primes in order to create interesting new music. There have also been a few attempts to transform aspects of the Riemann zeta function into (not particularly musical) sound. Due to its inherent irregularity, the beginning of the sequence of prime numbers can also be used to generate unusual rhythmic patterns. But in all of these undertakings, there are numerous variables and choices involved, so there can be no one true "music of the primes".

The seeming "importance" (to certain people) of the Riemann Hypothesis is arguably in the same category as the often claimed "importance" of celebrated Western music such as Beethoven's symphonies. What might appear to an entirely non-musical extraterrestrial as a horde of humans blowing into, scraping and banging on a curious array of strangely shaped objects can produce vibrations in the air, the effects of which on other nearby humans, it is claimed, can amount to something of real (almost cosmic) significance. Analytic number theorists can similarly claim that whether or not a bunch of seemingly obscure dots all lie on a vertical line amounts to something of real (almost cosmic) significance. However, many people "feel" the significance of music because it's immediately accessible to them in a way the Riemann zeta function is not. But keeping in mind some of the ideas discussed in recent chapters, we see that both music and the mysterious distribution of prime numbers could be seen as communicating (in different ways, perhaps to different parts of the mind) something of the mystery of time.

If we ask "Why do humans pursue a proof of the Riemann Hypothesis?", the question has a similar feel to "Why do humans compose and perform music?" Almost anyone involved in number theory or music would find it difficult to give a completely satisfactory answer, yet within themself there would be no sense of doubt regarding these matters. Both certain theorems of number theory and certain pieces of music are widely acknowledged as beautiful. Both have an ambiguous relationship with the world of physical matter. And, I would argue, both are relevant to the mysterious interrelation of time and number.

Chapter 37
names, numbers, archetypes

In the remaining chapters, we're going to return to some issues raised in Chapter 1: how we relate psychologically to number concepts and the role of the number system in how the Western version of "reality" is structured.

Once more we'll consider these words:

> "As archetypes of our representation of the world, numbers form, in the strongest sense, part of ourselves, to such an extent that it can legitimately be asked whether the subject of study of arithmetic is not the human mind itself. From this a strange fascination arises: how can it be that these numbers, which lie so deeply within ourselves, also give rise to such formidable enigmas? Among all these mysteries, that of the prime numbers is undoubtedly the most ancient and most resistant." (Gérald Tenenbaum and Michel Mendès France) [1]

I've quoted this before, in both of the previous volumes. I keep returning to it because, of everything I've read on the matter, it comes the closest to articulating what I feel is the key issue underlying everything dealt with in this trilogy.

So let's now take it apart, keeping in mind all that's been discussed up to this point:

> "As archetypes of our representation of the world, numbers…"

Numbers, the authors are saying, have some significant role ("as archetypes") in how we "represent" the world. The assumption here is that there's an objective world beyond the mind and that in order to interact with this we create a mental "representation" of it. Such a representation will depend on our cultural background, among other factors. By "*our representation*", the authors might have intended to include all of humanity. I'd be cautious, though, and take it as shorthand for "those of us humans who have been brought up in or heavily exposed to certain cultural groups (including Western cultures)". After all, there are cultures where numbers have a very different kind of role in the shared "representation" of the world, even if this might be seen as of marginal significance by some in the West. But as a modern Westerner, that "our" definitely includes me, and considering that you've been able to read this book, it almost certainly includes you as well.

So the authors are saying that counting numbers play a key role in how our representation of the world works – as "archetypes". The word "archetype" isn't easily defined. It continues to be used in a variety of ways by people with very different ideas about what it signifies [2]. This isn't really the place to discuss that, though (we'll come back to archetypes later). The best I can do here is to interpret what I believe the authors are trying to express.

We represent the world largely as a collection of objects (we'll have to use that word in a *very* general sense) and relationships between those objects. Very importantly, we group these objects together into categories, depending on what they "are", some examples being people, walruses, paperclips, acorns, bubbles, reggae bands, letters of the Cyrillic alphabet, bees, machines, poems by Ted Hughes, species of lichen, jokes, flavours, sequins, mathematical theorems, hippies, clouds, criminals, adjectives, silly noises, children and feelings.

As soon as we introduce categories, the number system becomes involved in our representation of the world. If everything you ever encountered was just "its own thing", "what it is" or "part of an inseparable unity" (the sort of insight people report having experienced in various mystical states of consciousness), then there would be no place for the number system. But as soon as you create a category – "person", "tree",

$$+\sum_{\{R\}_\Gamma}\sum_{k=1}^{m-1}\frac{\sigma(\chi^k(R))}{M\sin k\pi/m} \qquad \int_{-\infty}^{\infty}\frac{e}{1+e^-}$$

$$+2\sum_{\{P\}_\Gamma}\sum_{k=1}^{\infty}-\frac{\sigma(\chi^k(P))\log N\{P\}}{(N\{P\})^{k/2}-(N\{P\})^{-k/2}}$$

Big green

Splendid CREPUSCULAR

fluffy

"rock", "cow",... – then there can be two people, trees, rocks, cows,.... Or three, or four, or five,.... Suddenly you have the sequence of counting numbers.

Suppose I were to make a pile of seven beans, another pile with seven coins, another with seven matchsticks, one with seven pebbles and finally a pile with seven keys. All of these piles have something in common, which we could call their "sevenness". Clearly, the number 7 is somehow suggested by each pile.

One implication of this is that we can put the contents of any two piles in direct correspondence (pairing off beans with keys, or pebbles with coins).

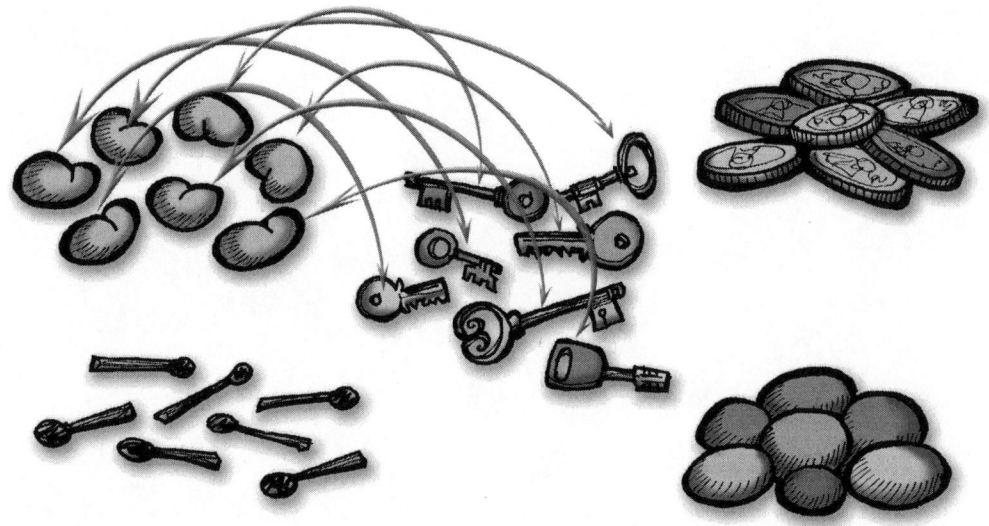

The contents of each pile can also be put in correspondence with the 1's that make up 7 in the Peano-based 1 + 1 + 1 + 1 + 1 + 1 + 1 approach (see Chapter 2). But we can't make a physical pile of 1's on the table. It has to exist *in our mind*. All of us who have learned to count, then, have these "archetypal" piles in our minds – they're what we call "numbers"! We put them in correspondence with the piles of matchsticks, beans, coins, keys and pebbles through the act of counting, like this:

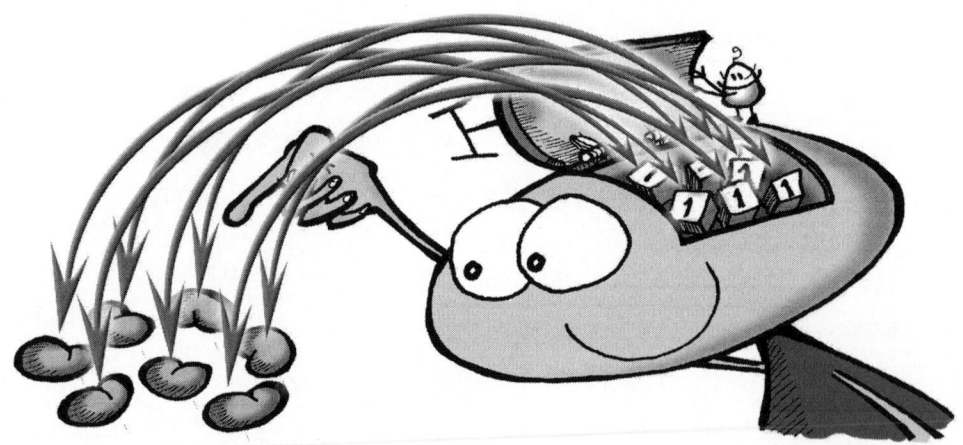

In this way, you ought to be able to see how the number 7 is a kind of universal "archetype" which is manifested physically by each of the piles.

With simple piles of beans, *etc.* there's very little room for disagreement. But generally, in order to come to an agreement as to whether an object should or shouldn't be classified as a "person", "walrus", "paperclip", "acorn", "bubble", or whatever category we're considering, we must have access to some kind of discernment. Is that object in question a frog or a toad? A tangerine or a clementine? A cup or a mug? A chair or a stool? A bee or a wasp? A reggae band or a ska band? A poem by Ted Hughes or a poem by someone else in a similar style? To answer these kinds of questions requires the ability to classify objects according to a combination of perception, recognition, memory, knowledge and judgment. In some cases the answer is never in serious dispute. In others it can become the subject of a controversial legal ruling. Often we must accept that it will depend very much on the individual. Is that a silly noise or a sensible noise? A funny joke or an unfunny joke? A good song or a terrible song?

The point here is that we group the "objects" of the world into categories, *but we all do this in slightly different ways because the categories don't have clear boundaries*. Generally this isn't a problem, but when we look closely at how this tendency to *categorise* objects relates to our tendency to *count* objects, things begin to get interesting.

Back in Chapter 1 I wrote:

"By *counting*, I mean the application of number to the physical world, by means of *agreed-upon categories of things-to-be-counted*.

Eh?

OK, try this: look around you and *count everything you can see*.

It's not so obvious is it? No, in order for counting to be meaningful, there must be an agreement as to a "category of thing" which you're going to count..."

Suppose I were to gather a few serious minded people in a kitchen and ask them to count all the cups. We might get different answers from different people, as some may count a particular object as a cup, while others may not, considering it to be a mug. Suppose, instead, I ask them to count all the "vessels". Or all the "utensils". These are valid categories, but their boundaries are even blurrier, so we can expect a wide range of answers depending on the interpretations of the people doing the counting.

This simple thought experiment illustrates that although we represent the world via the classification of objects into categories (which then allows counting), it's all underpinned by an invisible web of consensus as to how the categories are defined and how definitions are interpreted.

There's a striking contrast between the rigid, unchanging, "indisputable" nature of the number system and the much more vague, culturally dependent way in which we "map" our experience of reality. But they're strongly linked, because *as soon as you can name, you can count*, and if you're counting, *there has to be some nameable category-of-thing to count*.

In certain approaches to transpersonal psychology[3], "archetypes" are understood to be timeless, unchanging forms which emerge from the "collective unconscious" and manifest in the more fluid, somewhat arbitrary realms of mythology, folklore, culture, *etc.* (see Appendix 16). It's in that general spirit that we can consider counting numbers to be "*archetypes of our representation of the world*".

The Journey

Back to the Tenebaum and Mendès France quote we're analysing:

"As archetypes of our representation of the world, numbers form, in the strongest sense, part of ourselves…"

The concept of "self" is a problematic one. But as numbers seem to inhabit some shared level of mind, something we all have access to and can agree about, in that sense they're part of my "self", yours and (potentially) everyone else's. Now, a lot of what I casually consider to be my "self" is actually part of my *personality*, something which can change over time. Memories can be lost or distorted. Traumas and triumphs can significantly alter personality structures. So if I want to claim something as "part of my self", I must consider the extent to which it's changeable or transitory. These deep questions about the true nature of the self are what many contemplative and meditative traditions have been confronting for millennia. The quest for the "true self" involves delving ever deeper, discarding anything which is ephemeral. Numbers seem to persist "as deep into the mind" as anything else we deal with in our ordinary lives. In this way, we could argue that they "*form…part of ourselves*" in a very strong sense indeed.

"…numbers form, in the strongest sense, part of ourselves, to such an extent that it can legitimately be asked whether the subject of study of arithmetic is not the human mind itself."

I would argue that we don't really know what numbers are, but it seems that they're somehow part of our selves, our minds. So when we study number theory, the

workings of the number system, the interrelations between the counting numbers, *what are we actually studying?* The subject matter of zoology is animals, the subject matter of botany is plants, the subject matter of geology is rock formations, *etc.*, so what's the subject matter of number theory? Numbers of course! But unlike animals, vegetables or minerals, these are not only non-physical, not only "just in our minds", *they're a central part of how we represent the world and, therefore, how we think.* So when contemplating questions of number theory, we're effectively thinking about an archetypal structure which we depend on to think. Bearing this in mind, it's not surprising what a difficult subject it can be.

When the authors wrote *"the human mind itself"* (in French), they were perhaps being a bit vague. Psychology is the discipline whose subject of study is the human mind, and they weren't proposing that number theory and psychology are somehow the same thing. To be more precise, I would suggest "the subject of study of arithmetic is the deep, shared, 'representational' level of the human mind", although this still contains some (unavoidable) vagueness.

> *"…From this a strange fascination arises: how can it be that these numbers, which lie so deeply within ourselves, also give rise to such formidable enigmas? Among all these mysteries, that of the prime numbers is undoubtedly the most ancient and most resistant."*

As I explained in Chapter 3, the number system only becomes a complicated structure once we introduce multiplication[4]. From then on, we have access to the primes and

The Journey

the various mysterious issues that accompany them. These mysteries are the "*most resistant*" of the "*formidable enigmas*" to which the authors here refer.

The introduction of multiplication involves making a transition from just counting "objects" to also being able to count *numbers themselves* ("three sixes", "twelve twelves"). The number system becomes complicated at this point where numbers get "objectified", that is, recast as the sorts of things you can count. So although I'm not attempting to give a complete answer to the authors' question...

> "*...how can it be that these numbers, which lie so deeply within ourselves, also give rise to such formidable enigmas?*"

...I would propose that the number system, which lies "*so deeply within ourselves*", is able to "*give rise to such formidable enigmas*" because *we have turned it in on itself by objectifying numbers*, that is, by allowing numbers to become objects of counting (which is the essence of multiplication, which then gives us divisibility, the primes, the zeta function, its zeros, the Riemann Hypothesis, the spectral interpretation and all that goes with it).

MATHEMATICAL FEEDBACK

Counting numbers, it could be argued, are *concepts*, each counting number being a different concept. As discussed in Chapter 1, it was quite common in pre-Westernised cultures for individual numbers to be in some sense personified, deified or revered.

Each number had its own qualities, its own feeling, its own "vibe". This is the "qualitative" approach to number which was explored by the psychologist Carl Jung (and later by his student Marie-Louise von Franz[5]) and which is generally dismissed as meaningless or irrelevant by modern scientific culture.

By creating the category "counting numbers", a category containing an infinity of distinct concepts – "one", "two", "three", "four",… – Western civilisation made a kind of conceptual leap which both (1) began the process of the quantitative view of number eclipsing the qualitative view, and (2) allowed mathematics as we know it to be developed. This might help to explain Westernised humanity's peculiar relationship with the number system. From this perspective, the mystery of the primes, the zeta function, the Riemann Hypothesis, *etc.* can be understood as a kind of "feedback" caused by the objectification of number and the related "turning in on itself" of the number system.

Feedback is encountered in many forms, perhaps the most familiar being the high-pitched screech produced when a microphone is pointed towards an amplifier which it's connected to. The (usually) unintended noise is the result of the microphone picking up the sound of the amplifier so that the amplifier begins to amplify the sound of itself (a *feedback loop*). The intended function of the microphone is to pick up the sound of a human voice or musical instrument so that it can be amplified. So when the microphone picks up the sound of the amplifier, something has gone wrong, and feedback (in the form of a wavering, piercing noise) is the result.

The Journey

A similar kind of thing occurs when a video camera is pointed at a monitor to which it's connected. An endless splurge of morphing blobs and geometric forms appears on the screen, the result of the screen displaying an image of itself (with a small time delay and some image distortion unavoidably involved in the process). This is called *video feedback*, a particular form of *optical feedback*.

When they first surfaced consciously in human cultures, counting numbers were "intended" for use in the counting of objects – "discrete countable objects" as I called them in Chapter 35. The categorisation of reality's contents into these groups of corresponding discrete countable objects is a huge part of how we structure our perception ("look Mummy, an aeroplane!"). But when we use counting numbers to count *other numbers* ("three sevens", "twelve twelves"), the "microphone" is being used in a new, unintended way and causing the "amplifier" to "amplify its own sound" – the number system has been turned on itself and an endless torrent of "mathematical feedback" is created, which (at this point in history) is called "analytic number theory" and takes this kind of form [6]:

Similarly, the Lerch zeta-function (for its theory see e.g the monograph of Garunkštis-Laurinčikas [7]) defined by

$$\phi(\alpha,\beta;s) = \sum_{n=0}^{\infty} e^{2\pi i n\alpha}(n+\beta)^{-s}, \qquad (0 < \alpha, \beta \le 1)$$

for $\mathrm{Re}(s) > 1$ and by analytic continuation elsewhere, is continuous up to $\mathrm{Re}(s) = 1$ except for a possible pole at $s = 1$. Thus Lemma 15 also implies a similar theorem for this case:

Theorem 28. *For the Lerch zeta-function $\phi(\alpha,\beta;s)$ and $0 < \alpha, \beta \le 1$ we have the following result*

$$\beta^{-1}\left(1+\frac{\beta}{\delta}\right)^{-\frac{1}{14}}10^{-\frac{5}{2}} \le \inf_{T}\int_{T}^{T+\delta} |\phi(\alpha,\beta;1+it)|dt.$$

We will also state versions of these inequalities when we take the infimum of α and β. First we state two variants Lemma 15. Since the classical Dirichlet series case is of special interest we state a version for this case:

Lemma 16. *Assume that $|a_n| \le 1$. Then we have for $0 < \delta \le 0.05$ that*

$$\inf_{\sigma>1,T}\int_{T}^{T+\delta}\left|1+\sum_{n=1}^{\infty}a_n n^{-\sigma-it}\right|dt \ge \delta^{\frac{7}{14}}10^{-\frac{5}{2}}.$$

Proof. This is proved in the same way as Lemma 15, and follows from the error thrown away when using (34) and (35). In particular, in Lemma 15 we could have stated the right hand side as

LATTICES AND VEILS

In 1963, Aldous Huxley was introducing some "new" (old!) ideas into mainstream Western culture, this being a precursor to the widespread interest in Eastern mysticism that came out of the Western psychedelic counterculture in the latter half of that decade:

"*Trobriand Islander or Bostonian, Sicilian Catholic or Japanese Buddhist, each of us is born into some culture and passes his life within its confines. Between every human consciousness and the rest of the world stands an invisible fence, a network of traditional thinking-and-feeling patterns, of secondhand notions that have turned into axioms, of ancient slogans revered as divine revelations. What we see through the meshes of this net is never, of course, the unknowable 'thing in itself'. It is not even, in most cases, the thing as it impinges upon our senses and as our organism spontaneously reacts to it. What we ordinarily take in and respond to is a curious mixture of immediate experience with culturally conditioned symbols, of sense impressions with preconceived ideas about the nature of things. And by most people the symbolic elements in this cocktail of awareness are felt to be more important than the elements contributed by immediate experience. Inevitably so, for, to those who accept their culture totally and uncritically, words in the familiar language do not stand (however inadequately) for things. On the contrary, things stand for familiar words. Each unique event of their ongoing life is instantly and automatically classified as yet another concrete illustration of one of the verbalised, culture-hallowed abstractions drummed into their heads by childhood conditioning.*"[7]

"*The universe in which men pass their lives is the creation of what Indian philosophy calls* Nama-Rupa, *Name and Form. Reality is a continuum, a fathomlessly mysterious and infinite Something, whose outward aspect is what we call Matter and whose inwardness is what we call Mind. Language is a device for taking the mystery out of Reality and making it amenable to human comprehension and manipulation. Acculturated man breaks up the continuum, attaches labels to a few fragments, projects the labels into the outside world and thus creates for himself an all-too-human universe of separate objects, each of which is merely the embodiment of a name, a particular illustration of some traditional abstraction. What we perceive takes on the conceptual lattice through which it has been filtered. Pure receptivity is difficult because man's normal waking consciousness is always culturally conditioned. But normal waking consciousness, as William James pointed out many years ago, 'is but one type of consciousness, while all about it, parted from it by the filmiest of screens, there lie potential forms of consciousness entirely different. We may go through life without suspecting their existence; but apply the requisite stimulus, and at a touch, they are there in all their completeness, definite types of mentality which probably somewhere have their field of application and adaptation. No account of the universe in its totality can be final which leaves these forms of consciousness disregarded.'*"[8]

At the end of Chapter 22 I described the early Buddhist concept of *jñeyāvarana* or the "veil of fixed ideas" [9], the establishment of which begins with the basic "is/is not" distinction-making of *prapañca* or the "multiplicity of named things". The latter appears to be similar to the "Nama-Rupa" which Huxley refers to. The "conceptual lattice" which he describes raw "Reality" being filtered through, and our perception being shaped by, will be culturally dependent, though: a traditional Trobriand Islander may well inherit a conceptual lattice from his or her enculturation process which differs sharply from that of a suburban Bostonian. The deepest layers of *jñeyāvarana*, as I understand it, lie within the shared levels of mind. Whereas anthropology could be said to concern itself with various cultures' "conceptual lattices", number theory, I would suggest, is dealing with the deep architecture of the "veil", which is the same for everyone. For as soon as you have a "multiplicity of named things" (whatever the names or things) you have the possibility of counting, of number, of the numbers themselves becoming "named things"– hence of multiplication, divisibility and everything that comes with the distribution of prime numbers. The curious "shape" that this takes could be a reflection of the "shape" of structures in shared levels of mind which are responsible for producing multiplicities of named things. It's in this sense that I understand Tenenbaum and Mendès France's question as to "*whether the subject of study of arithmetic is not the human mind itself.*"

SHARED LEVELS OF MIND

The idea of numbers being "*part of ourselves*" might suggest to those with a more scientific-materialist mindset that the number system is somehow hard-wired into the

The Journey

structure of the human brain. This would account for the way we all have access to it and the fact that children (with almost no exceptions) have very little trouble "getting" the basic ideas of counting and addition. However, as mentioned in Chapter 35, recent research suggests that it's only the first few counting numbers which are hard-wired into the brain. So if the number system *as a whole* is "part of ourselves", a more plausible possibility is that it's somehow embedded in a shared level of *mind*.

Carl Jung was very much concerned with "shared levels of mind". He introduced the notion of the *collective unconscious* as well as the idea of archetypes inhabiting it. Towards the end of his life, he became particularly interested in the archetypal role of numbers. This came more from his observations about certain counting numbers recurring in particular contexts in fairytales, folklore, religious texts and imagery worldwide than from any substantial knowledge of number theory. Although this aspect of his work was never fully developed, by the time of his death he'd arrived at the idea that the system of counting numbers was a manifestation of the "archetype of order", a sort of "super-archetype" which ordered (in the sense of "arranged" or "configured") all of the other archetypes.

Jung also introduced the concept of the *unus mundus* (also known as the *psychoid*). This is the mysterious domain which supposedly exists between, bridges or transcends the realms of mind and matter, where the collective unconscious is to be found. (A glossary of basic Jungian terminology can be found in Appendix 16.)

Jungian theory has a substantial following, but there are also many psychologists in

The Journey

2013 who take issue with aspects of it. It's not my aim to promote Jung's ideas, but many issues which I want to discuss in relation to the number system almost seem to be inviting a Jungian interpretation. We'll see more of this in Chapter 41.

Tenenbaum and Mendès France, two accomplished mathematicians, have suggested that the number system is part of ourselves, part of our (shared) mind. Yet number clearly manifests in the physical world – when looking at a honeycomb or cluster of quartz crystals, it's hard to dispute the inherent "sixness" of the hexagons you see, and when considering a starfish or five-petalled wildflower, it's hard to dispute the inherent "fiveness". The number system seems to have one foot in material reality and the other in "psychic reality". It's acting as something like a bridge between the realms of psyche and matter, which places it squarely in Jung's *unus mundus*.

The unconscious mind (both collective and personal) is often symbolised by the ocean – something with tremendous depths, often unexplored, containing creatures which we can find alien or disturbing. A metaphorical image which this brings to mind is that of the number system as a sea creature which Western civilisation has "caught" and dragged onto the dry land of the conscious mind, to be examined in broad daylight. Its natural home would be in the unconscious, although it's known to stick its head above the surface (almost all human cultures consciously deal with counting numbers to some extent). For the great bulk of human history it's been an unconscious entity with some elements emerging into consciousness. Human cultures have generally been aware of the existence of this creature, but have tended to leave it alone, respectfully allowing it just to be.

The project initiated by the ancient Greeks (arguably with earlier roots in Egypt) and continued worldwide today by a global community of academics, what we know as "mathematical research", has involved the capture and dissection of this creature in an attempt to study its anatomy and understand "how it works".

I should point out that the mathematical community tends to have very little interest in the sorts of issues I've been discussing in this chapter. There are exceptions of course, but because it's perfectly possible to carry out detailed research, prove theorems, *etc.* without being concerned with "what mathematics *is*" or "what it *means*", and because this research can be extremely time-consuming, requiring intense dedication and focus, it's understandable that questions which are not strictly part of mathematics, but rather part of philosophy, are left to "philosophers of mathematics".

chapter 38

quantocentrism

Having been concerned with the subject for almost my entire life, but generally moving in social circles outside the world of academic mathematics, I've noticed how often people, when it comes up in conversation, will declare something like "I hate maths!". This seems to be very common in modern Western culture.

A consequence of this widespread "hatred" is that in many social scenes, a person admitting to an enthusiasm (or even competence) for mathematics risks being ostracised. The dominant trend in early 21st century Western society (the most "mathematicised" in human history) is to happily admit to, even revel in, a distaste for the subject. Number concepts have become linked to negative emotions in a large proportion of the population. Does this tell us something about the current status of number within the "collective psyche" of Western society?

The cause of these emotions is no great mystery. It's almost entirely the result of people's experiences at school, being forced to study something they couldn't understand and/or couldn't see the point of studying. To "drive" the modern world, that is, to maintain economic growth, Western (and Westernised) societies need a certain small proportion of the population to be mathematically competent at various levels in order to be the economists, financiers, engineers, statisticians, *etc*. Current mathematics education is primarily designed to fulfil these large-scale economic needs. From this point of view, everyone who isn't going to become part of this small "numerate elite" is wasting their time learning mathematics beyond the basics needed to function in society. But rather than selectively educating those children who show a natural aptitude and enthusiasm for the subject, *everyone* is expected to learn it. The fact that such a large proportion of adults end up hating maths is seen as a minor unfortunate consequence which is given little thought.

When talking to people about their experiences in this area, I've noticed that the "I hate maths!" reaction is often followed by a qualification like "Well actually I quite *enjoyed* it, up until we had to do long division" or "...until we got to logarithms" or "...until we had to do that stuff with x's and y's" or "...until we had to solve quadratic equations". It seems that Western children generally enjoy learning mathematics when they can understand it, but once it surpasses their understanding (often due to the style of teaching) they're left feeling confused, mystified, incompetent or stupid in the face of something they're supposed to understand but can't – an obviously unpleasant experience. This is exacerbated by the fact that teachers are often unable to explain to them *why* they're learning these things. In the worst cases, it can seem almost sadistic, a kind of mental torture.

I've often thought of this process as being like a train. Everyone gets on the same train when they learn to count as an infant, and it sets off down the track, which represents the maths curriculum of the current education system. The train gradually accelerates, causing individual children to fall off at the point where the subject no longer makes sense to them. They're left beside the track, mentally bruised, cursing the subject and feeling stupid or incompetent for the remainder of their maths education. The train disappears over the horizon and they either struggle in vain to catch up with it, or just give up trying.

I'm not particularly concerned here with the reform of Western mathematics education. On that front, there are endless attempts (most of which make me cringe) to make the subject "fun" or "cool" for children. But there's almost never any questioning of *why they must be taught it* and very little consideration of how it might be affecting them psychologically.

So the Western world is full of adults who have undergone minor traumas *in relation to a body of abstract ideas*. And as far as I'm aware, no one is discussing this. Beyond the unfortunate consequences for the individuals involved, I'd like to raise the issue of how this might be affecting the *collective* psyche. It's a puzzling situation we find ourselves in. As discussed in Chapter 1, we've built our civilisation to a very great extent on number-based thinking, and the process of "mathematizing" just about everything is continuing to accelerate. At the same time, an alarming proportion of the population operating within this framework (I suspect a considerable majority) has a negative emotional relationship with number-related concepts, even if this is something that's rarely given any thought.

If number is as deeply embedded in the collective psyche as some of the observations and ideas I've presented would suggest, then this widespread emotional stance may turn out to be of real significance for the collective mental health of society. This is something which, to my knowledge, has never been given any serious consideration.

The Journey

What can be done about this? Presenting number concepts to children in a way designed to cultivate a sense of wonder might produce generations of mathematically aware adults who don't necessarily want to participate in the ongoing process of further "mathematizing" everything around them. They might even question the dominance of economics and technology. Ambitions to become economists, engineers and accountants might be displaced by a preference for simply contemplating the mysteries of the number system and the wonders of creation. However much this could improve the collective mental health of society, though, it wouldn't promote economic growth, so it's currently out of the question.

In recent years there's been a noticeable flourishing of popular interest in the less practical, more "beautiful" areas of maths – fractals, for example, and what often gets called "sacred geometry"[1]. This has had much to do with the New Age movement which (at least in its pre-commercialised form) runs against the grain of the increasingly materialistic and "mathematized" nature of society. Although this phenomenon is still relatively marginal, it's interesting to imagine what the social effect of mass exposure to such mathematical ideas might result in.

A POSSIBLE DIAGNOSIS

One view of itself which Western culture continues to promote is that by mastering mathematics and the sciences it was able to develop technologies and thereby rise above, and then bring civilisation to, the rest of humanity.

Here's a different view which might be worth considering: The Western collective psyche, through its curiously unbalanced relationship with the number system, has been somehow disturbed to the point that Western humanity has fallen out of harmony with the "whole" of which it is a part[2]. It's been steadily losing its ability to relate to the non-material and unquantifiable aspects of the world. This imbalance has caused it to rampage around the planet, conquering and enslaving other peoples, stealing resources, destroying cultures, wiping out species, polluting ecosystems and quantifying everything in its path. It would be arrogant to argue that this behaviour was simply the result of "human nature", for that would imply that other cultures haven't acted like this (at least not on anything like such a vast scale[3]) simply because they weren't intelligent or hard-working enough. Rather, it seems to me, this is probably because it wasn't in their nature[4].

It's like a "socio-cultural disease" with which the West has now infected almost all other cultures on the planet. They're all now tied into the global economy, and their collective decision making[5] is dominated by quantity-based considerations far more than it was before contact with Europeans. The unquantifiable and non-material aspects of their cultures, which had previously been of central importance to them (folk wisdom, folklore, a subtle network of community bonds, music, ancient languages, notions of "spirit" and other realms of being, ancestral awareness, *etc.*) are generally fading away, being replaced by an increasingly uniform worldwide Western-style consumerism.

By using the word "disease", it might appear that I'm making a value judgment that Western culture is "bad" or "wrong", but that really isn't my intention. I've chosen this analogy with some care. A disease is not "bad" or "wrong". It has no evil intentions – or any intentions at all (beyond, in the case of a disease-causing virus, reproducing itself). It's just "doing its thing", being a disease. Still, we could certainly agree that *from our body's point of view*, a disease is undesirable. It's in no way beneficial and is potentially disabling or damaging to the body's workings. A disease creates instability in an otherwise stable body.

If you look at the "whole" of which humanity is a part, what's sometimes called the *biosphere*, it looks a lot less healthy and stable than it would have done, say 10 000 years ago [6]. So the source of this damage or instability could reasonably be compared to a disease in the context of the biosphere. Humans are undeniably involved, with some groups clearly more deeply implicated than others.

Although my education trained me to think "scientifically", since being exposed to the writings of Marie-Louise von Franz [7] and René Guénon [8] (among others) I've arrived at a somewhat different perspective. But rather than simply dismissing the quantitative approach to number (as the Western mainstream has done with the qualitative), I began to think, as these authors suggest, that the West has fallen into a kind of inappropriate relationship with number concepts. To me, this goes a long way towards explaining its unusual behaviour in recent centuries.

There has been so little discussion of these matters that even to describe the outward manifestation of this (collective, inner) relationship with number almost requires a new word. I'd recommend *quantocentrism* [9], that is, a tendency to see the world exclusively in terms of quantities (and that includes all forms of measurement).

I think it's fair to say that Westernised culture distinguishes itself from all others in the following two ways:

(1) the extent of its rejection of the qualitative aspect of number [10] in favour of the quantitative (this can be related to its success in developing mathematics and quantitative science);

The Journey

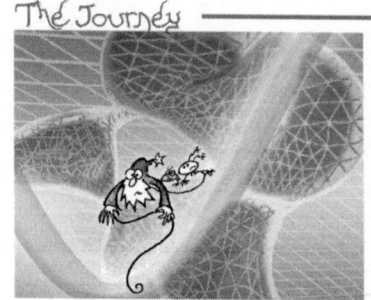 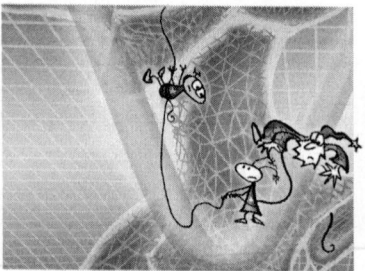 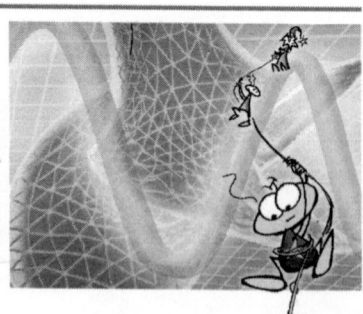

(2) the extent to which it has become fixated on number and quantity – not just the acquisition of vast quantities of resources by any means necessary, but also the project of trying to quantify everything, imposing a global economic system and promoting quantitative science as the only valid route to truth (among the consequences of this have been a worldwide trend towards various forms of standardisation and the mediation of ever more aspects of reality via digital technologies).

Could it be that these two phenomena are related?

Many books have been written attempting to diagnose "what has gone wrong" with humanity, attempting to relate it to settled versus nomadic life, agriculture, diet, patriarchy, usury, alphabetic languages[11], sunspot cycles, astrological forces and extraterrestrial interference, among other things. These authors (including myself) can't all be right! And nothing is ever as simple as we might like, so if my own obscure attempt at a diagnosis ("quantocentrism", due to an inappropriate relationship with number) has any truth in it, it's probably only going to be a partial truth. Also, I'm not even beginning to address the questions of *how* or *why* Western civilisation fell into this inappropriate relationship with number.

My simplified version of the story is that a certain sector of humanity dragged the number system from the oceanic unconscious into the full light of consciousness. It fell under the spell of number. It then began to quantify and commodify everything until it had commodified *number itself*. In the process, it lost touch with the qualitative aspect of number, eventually ridiculing the notion that there could even be such a thing.

Economics, science and technology, number's offspring, began to dominate its world. It felt driven to subjugate all other peoples who didn't see things in the same way and to impose upon them a life increasingly based on quantity. Its religion was a religion of number, of quantity. *But it didn't realise this* – its previous religion had gradually lost influence and it had started to believe that it no longer had any need for religion. It began to destroy its surroundings and poison itself in pursuit of ever larger quantities. Anything that couldn't be measured or quantified no longer mattered, so it couldn't see any reason not to continue like this (despite the consequences being clear to see to anyone outside the "quantocentric" way of thinking).

As well as an inappropriate relationship with number, I think it's reasonable to suggest that Westernised cultures have an inappropriate relationship with *time*. This is evident in Western attitudes towards aging and death, its continual obsession with "the latest thing", the widely-observed tendency for just about everything to speed up and the uniquely Western phenomenon of the *zeitgeist*[12]. Considering the links between number and time discussed in Chapters 34–36, it seems plausible that there might be some connection between this pair of inappropriate relationships.

HOPE?

This might come across as a somewhat gloomy assessment of things, but from my perspective in 2013, I can see a ray of hope. Based on the discoveries relating number theory to physics which were described in the earlier chapters of this volume, I put

The Journey

forward the idea that *the number system is something very different from what we had thought it was*. These discoveries may force us (or at least the "inner priesthood" of the mathematical-scientific establishment) to reassess what the number system "really is". Any major shift in perspective could then filter into a wider field of awareness, into the collective psyche, gradually altering our inner relationship with number. Eventually, such a process might just counteract the "quantocentrism" I've described and lead to a healthier, more harmonious and sustainable existence on this planet. I strongly sense that we're on something of an historical cusp in regards to our relationship with the number system. However, it's impossible to know what sort of time scale is involved, so whether something like this could ever have any impact on the multiple urgent crises currently facing humanity remains to be seen.

Chapter 39

Surprise!

There's a theme running through many of the ideas we've looked at in this trilogy, one which could perhaps be best summed up in the word *contradiction*. "Contradiction" has a formal meaning in mathematics and logic, but I'm using it in a looser, more colloquial sense because this isn't about mathematics itself, but rather about people's reactions to certain mathematical ideas. The latter is a potential field of study which hasn't really been developed (as of 2013), so it's not surprising that there's never been any substantial discussion of this "contradiction" phenomenon which we're about to explore.

The words *paradox* and *enigma* also convey some of the flavour of what I'm trying to convey here. Like "contradiction", "paradox" has a precise meaning in the study of logic, but I'm using it in a similarly loose, colloquial sense. An enigma can be thought of as a kind of riddle. There seems to be a particular brain function or aspect of mind that connects with riddles, puzzles and paradoxes (this would include Zen *kōans* as well as Sufi jokes and stories). When confronted by one of these, there's initially a kind of internal obstruction or conflict. Something appears to be two things at once which contradict each other. Then, when the solution appears, you "get it" and perhaps laugh, due to an involuntary muscular spasm, or, in the Zen and Sufi traditions, spontaneously achieve some sudden expansion of consciousness. The blockage is removed, the conflict is resolved and everything "flows" again.

The prime numbers present themselves to us as a colossal, ancient riddle. "How can we be *this* and yet be *that* at the same time?" they've asked generations of mathematicians. And still, no one has anything like a complete answer.

"To some extent the beauty of number theory seems to be related to the contradiction between the simplicity of the integers and the complicated structure of the primes, their building blocks. This has always attracted people." (Andreas Knauf) [1]

It could be argued that Knauf is misusing the word "contradiction" and that the word "contrast" would be more appropriate here. But using the word as he does, he's capturing something of the *feeling* that's induced by juxtaposing the Fundamental Theorem of Arithmetic (the fact that the primes act as building blocks of the counting numbers) and the irregularity of the distribution of primes within the system of counting numbers.

Humanity's long-standing inability to prove the Goldbach Conjecture illustrates a related contradiction. It's incredibly easy to state:

> *Every even counting number greater than 2 is the sum of two primes.*

But more than two centuries of research have seen very little progress towards a solution, and the current research typically looks like this [2]:

$$S_{n_0,1}(x/K,\alpha) = \frac{x}{K}(\hat{n}_0(\alpha x/K) + \mathcal{O}^*(10^{-6})).$$

By Lemma 4.1 and (8.13) (and bounding $\hat{n}_0(\alpha x)$ as $\mathcal{O}^*(1)$ whenever necessary) we then have

$$S_{\overline{n},\sqrt{x/K_1}}(x/K,\alpha)D_{N_0/3}(\alpha)^3 = \frac{x(N_0/3)^3}{K}(\hat{n}_0(\alpha x) + \mathcal{O}^*(1.1 \times 10^{-6})).$$

Let us consider the contribution of the error term $\frac{x(N_0/3)^3}{K}\mathcal{O}^*(1.1 \times 10^{-6})$ to (8.12). By Corollary 4.7, we may bound this contribution in magnitude by

$$1.1 \times 10^{-6}\frac{x(N_0/3)^3}{K}\frac{2}{1 - \frac{\log(2T_0/3x)}{\log x}}S^2_{\overline{n},\sqrt{x/K_1}}(x,0)$$

which by Lemma 4.3 and the bounds $T_0 = 3.29 \times 10^9$, $x \geqslant 8.7 \times 10^{36}$ can be safely bounded by

$$10^{-3}\frac{x^2(N_0/3)^3}{K}.$$

Thus it will suffice to show that

$$\int_{|\alpha|\leqslant\frac{T_0}{2\pi x}} S_{\overline{n},\sqrt{x}}(x,\alpha)^2\hat{n}_0(\alpha x/K)e(-x\alpha)\,d\alpha = x(\tfrac{2}{3} + \mathcal{O}^*(0.09)). \qquad (8.14)$$

We apply Proposition 7.2 again to obtain

$$S_{\overline{n},1}(x,\alpha) = x\hat{n}(\alpha x) + \mathcal{O}^*(A_1\frac{\log T_0}{2T}x + 2.01c_1^{-1/2}x^{1/2}N(T_0)\|\overline{n}\|_{L^1(\mathbb{R})})$$

There's no contradiction here in the formal sense, but it feels like there's something "contradictory" about this state of affairs.

184

A similar kind of "contradiction" exists between the indisputable, universal, "God-given" nature of the prime numbers and the fact that they're often described as being "disorderly", metaphors involving imagery like "weeds" and "jumbles" abounding. Related to this is the contradiction between the seeming randomness of individual prime number locations and the "military precision" of their collective behaviour.

Another such "contradiction" we encountered was between the "purer than pure" nature of prime number theory (within the wider framework of the mathematical sciences) and the fact that it has recently started to show up, unexpectedly, in numerous areas of physics.

To me, the common thread running through these "contradictions" can be summarised as follows: Something is a certain way, but we

✩ expect it to be...

✩ would like it to be...

✩ can't understand why it isn't...

✩ are annoyed by the fact that it isn't...

✩ are entirely unable to handle the fact that it isn't...

✩ had never even considered the possibility that it could be...

another way.

The Journey

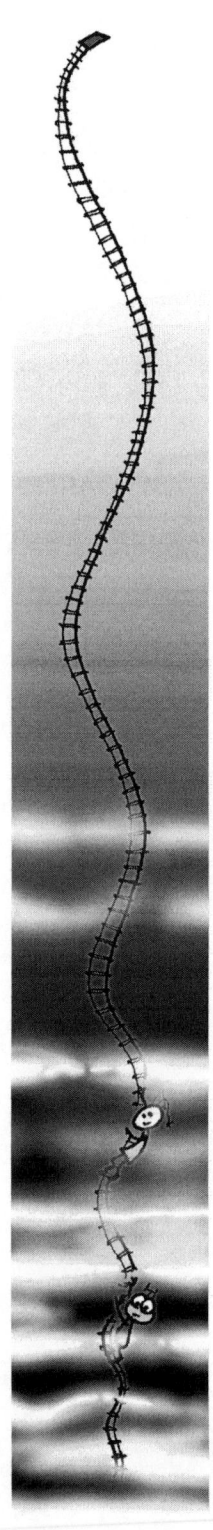

We've seen the repeated use of words such as "surprising", "astonishing", "baffling", "stultifying", "exasperating" and "stunning" in connection with certain prime-related issues. These all indicate variations on a certain state of mind which arises when confronted by something which "doesn't quite fit", or when something is other than we expect (or want) it to be.

Various "surprises" have shown up in this trilogy (obviously some are more surprising than others, this partly depending on the level of knowledge of the person encountering them):

✮ the Fundamental Theorem of Arithmetic (the fact that every counting number can be built from primes according to a unique "multiplicative" recipe) [Chapter 4]

✮ the random-looking sequence of factorisations (that is, the way the number of factors varies unpredictably from one counting number to the next) [Chapter 4]

✮ that it's possible to prove there are infinitely many prime numbers (and so easily!) [Chapter 7]

✮ the lack of any "pattern" (in the familiar sense) in the sequence of primes [Chapter 8]

✮ that there *is* a pattern of sorts, but one of a "statistical" nature [Chapter 10]

✮ that what I call the "primeness count deviation" splits cleanly into an infinite sum of harmonics [Chapter 13]

✮ that π shows up in connection with probabilities of "relative primeness" [Chapter 18]

✫ Euler's product formula [Chapter 18]

✫ the proliferation of diverse reformulations of the Riemann Hypothesis [Chapter 24]

✫ the connection between the Riemann zeta function and random matrix theory [Chapter 29]

✫ that the nontrivial Riemann zeta zeros appear to be the spectrum of an unknown "something" [Chapters 28–29]

✫ the link between the Riemann zeta function and quantum chaology [Chapter 30]

✫ that number theory is beginning to show up in numerous other areas of physics [Chapter 32]

✫ the extent to which randomness seems to play a role in the number system – that there's such a thing as "probabilistic number theory", with the primes behaving "*like experimental data*" [Chapter 33]

To number theorists, of course, a lot of this *isn't* surprising. Once you get used to the fact that the number system behaves in certain ways, it's no longer able to surprise you. It's the *initial impact* of these ideas on the psyche which is most revealing. Many number theorists may well have been surprised by these facts when first encountering them during their mathematical education, but the surprise is soon discarded and replaced by knowledge and familiarity.

The Journey

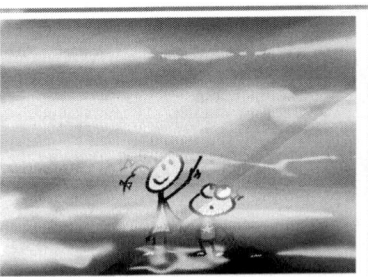

Probably most surprising to the mathematical community has been the connection with quantum chaology. Not only are the nontrivial Riemann zeta zeros behaving as if they were eigenvalues of something, it's a *very specific* something studied in an obscure branch of physics. Unlike the bulk of the surprises just listed, this takes us well outside the usual territory associated with number theory and therefore can't ever be accounted for by purely mathematical arguments. And it seems that we're a long way from understanding or explaining this connection (an explanation would largely "dissipate the surprise" for those who understood it).

The probabilistic or "stochastic" (almost "gas-like") nature of the primes revealed by the connections with statistical physics only serves to compound this surprise. Alongside the mysterious fact that the zeta zeros look exactly like the eigenvalues of something, there's the mysterious fact that they appear to be the singularities of a "partition function" associated with phase transitions arising in statistical mechanics (see page 75). That the zeta zeros should be behaving in either way is mysterious enough... but both simultaneously?!

Based on what we've seen, we have every reason to believe that the surprises will continue to emerge as we dig deeper and discover more. Here are some relevant quotations involving varying degrees of surprise (I've added all of the boldface emphases):

> "*What is **rather startling** is that this very regimented behaviour of the harmonics* [as predicted by the RH] *implies the cointossing nature of the primes.*" (Marcus du Sautoy)[3]

The Journey

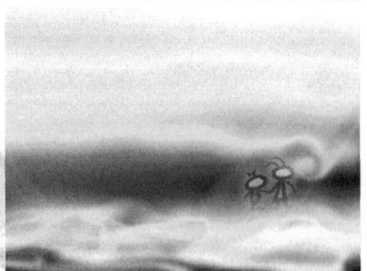

"*Much of mathematics may seem inscrutable, but the numbers 1, 2, 3, … spill out, known and familiar… [T]he understood aspect of the natural numbers is their additive structure. Getting from one number to another by addition or subtraction poses no mysteries. It is only when we start to think of things multiplicatively that the trouble starts and **the surprises enter**.*" (Dan Rockmore) [4]

"*To get a good understanding of how the Riemann hypothesis impinges on the law of distribution of primes, it is necessary to introduce a new tool, also described by Riemann, and known under the name of* explicit formulae. *The point here is to push to its natural limit the description of the* [prime counting] *function… in terms of the zeros of $\zeta(s)$. That such an enterprise should even be possible **may be surprising**.*" (Gérald Tenenbaum and Michel Mendès France) [5]

"*Berry and his collaborator Jon Keating used* [the correlation between the spacings of Riemann zeros and energy levels] *to show how techniques in number theory can be applied to problems in quantum chaos and vice versa. In itself such a connection is very tantalising. Although sometimes described as the Queen of mathematics, number theory is often thought of as pretty useless, so this deep connection with physics is **quite astonishing**.*" (Julian Brown) [6]

"*That there exist such similarities between the studies of the semiclassical limit of quantum mechanics and the theory of numbers is **truly surprising**. **Even more remarkably**, similar stories can be told for many other areas of theoretical physics. Indeed there are now conferences with titles like 'Number theory and physics'.*" (Jonathan Keating) [7]

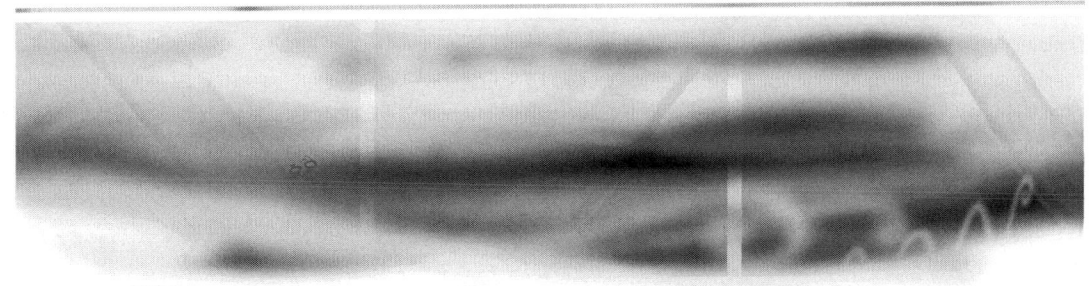

"*...in one of those* **unexpected connections** *that make theoretical physics so delightful, the quantum chaology of spectra turns out to be deeply connected to the arithmetic of prime numbers, through the celebrated zeros of the Riemann zeta function: the zeros mimic quantum energy levels of a classically chaotic system. The connection is not only deep but also tantalizing, since its basis is still obscure – though it has been fruitful both for mathematics and physics.*" (Michael Berry) [8]

"*The distribution of the individual primes among the integers is extremely irregular. But this irregularity 'in the small' disappears if we fix our attention on the average distribution of the primes...The simple law that governs the behaviour of this ratio is* **one of the most remarkable discoveries in the whole of mathematics**."

"*That the average behaviour of the prime number distribution can be described by the logarithmic function is a* **very remarkable discovery** *for* **it is surprising** *that two mathematical concepts which seem so unrelated should be in fact so intimately connected.*" (Richard Courant and Herbert Robbins) [9]

"*I hope that...I have communicated a certain impression of the immense beauty of the prime numbers and the* **endless surprises** *which they have in store for us.*"

"*For me, the smoothness with which this curve climbs is* **one of the most astonishing facts in mathematics**."

"*The second fact is* **even more astonishing**, *for it states just the opposite: that the prime numbers exhibit stunning regularity, that there are laws governing their behaviour, and that they obey these laws with almost military precision.*" (Don Zagier) [10]

"*The prime numbers are useful in analyzing problems concerning divisibility, and also are interesting in themselves because of some of the special properties which they*

The Journey

*possess as a class. These properties have fascinated mathematicians and others since ancient times, and the richness and beauty of the results of research in this field have been **astonishing**.*" (Donald Spencer)[11]

"[Prime numbers] *are **full of surprises** and very mysterious ... They are like things you can touch ... In mathematics most things are abstract, but I have some feeling that I can touch the primes, as if they are made of a really physical material. To me, the integers as a whole are like physical particles.*" (Yoichi Motohashi)[12]

"*It is a **pleasant surprise** that the Wiener–Khintchine formula which normally occurs in practical problems of brownian motion, electrical engineering and other applied areas of technology and statistical physics has a role in the behaviour of prime numbers which are studied by pure mathematicians.*" (H. Gopalkrishna Gadiyar and R. Padma)[13]

"*New approaches show promise, potentially bringing a proof of the Riemann hypothesis within reach ... One of the most exciting possibilities involves an **astonishing, unexpected connection** between the distribution of prime numbers and the energy levels of excited atoms.*" (Ivars Peterson)[14]

"*The zeta function plays **a surprising role** in number theory; for instance, the zeros of zeta are intimately connected to variations in the distribution of prime numbers.*" (Larry Carter)[15]

"[Euler's Product] *formula is fascinating because it converts a sum of powers to a product of prime numbers, and although it isn't difficult to prove, it is still a **surprising result**.*" (Steve Mayer)[16]

"*We exhibit a sequence c_n such that the convergence of $c_1 z + c_2 z^2 + c_3 z^3 + \cdots$ for $|z| < 1$ is equivalent to the RH. Numerical investigation of the c_n revealed some **astonishingly deceptive behaviour**.*" (Warren Smith)[17]

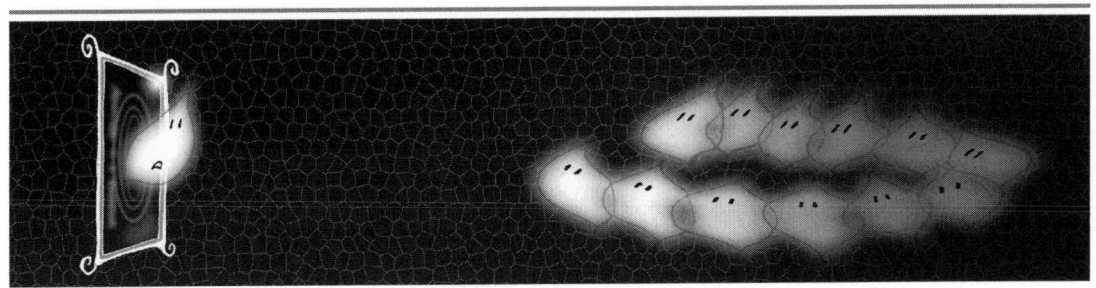

*"There exists a special mathematical model for the set of action integrals Φ, which is **almost too good to be true**: $\Phi = \log p$ where p runs through the set of prime numbers."*

*"**Most remarkably**, the mechanical interpretation has yielded a sharp mathematical result."* (Martin C. Gutzwiller) [18]

As I pointed out in Chapter 34, surprise occurs when reality differs from what you expect of it. When the 17th century scientist Antonie van Leeuwenhoek first looked through a microscope and saw microorganisms (or "animalcules" as he called them) he was greatly surprised. He hadn't even considered that such small animals could exist. These days, of course, we're not surprised by this at all. The light of science has illuminated the situation for all to see. In this sense, science could be thought of as a process, the ultimate aim of which is *to eliminate all surprises*.

Surprises can come in both pleasant and unpleasant forms. By expanding our scientific knowledge, we're able to predict what's going to happen in certain situations and, in this way, eliminate unpleasant surprises. Everyone likes pleasant surprises, but (by definition) you can't will them to happen – they just come along unexpectedly. You might have noticed how many of the expressions of surprise concerning the primes, zeta function, *etc.* suggested a kind of "delight". These are pleasant surprises to many people who encounter them. The number system reveals itself as being something different from what we expected, and rather than being horrified, we're delighted. Why should this be? Perhaps this tells us something about our relationship with the number system. It could be that these unexpected revelations hint that the number system (and with it, reality at large) is richer, weirder and a lot less uniform

The Journey

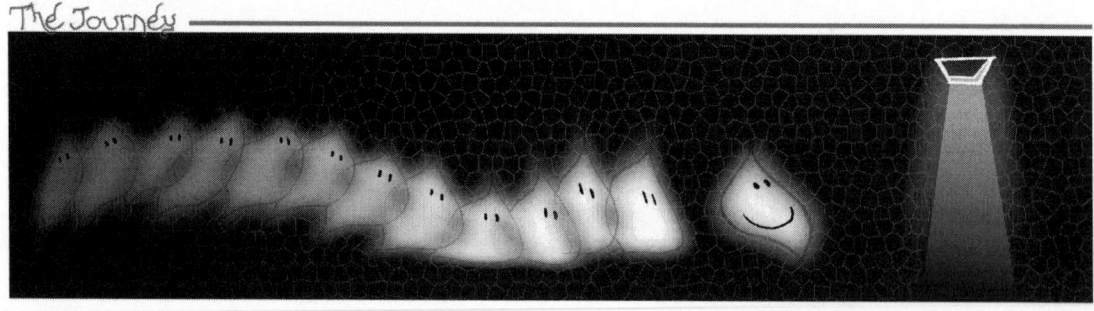

and mundane than we've been led to believe. Given enough popular exposure, this could potentially have the effect of counteracting the collective negative emotional relationship with the number system which I described in Chapter 1 and referred back to in Chapter 38. The appearance of fractal images in the 1980s and 90s had a similar effect of delighting many who first saw them, the pleasant surprise being that mathematics (associated with dull lessons at school, indecipherable equations, *etc.*) could do *this*: [19]

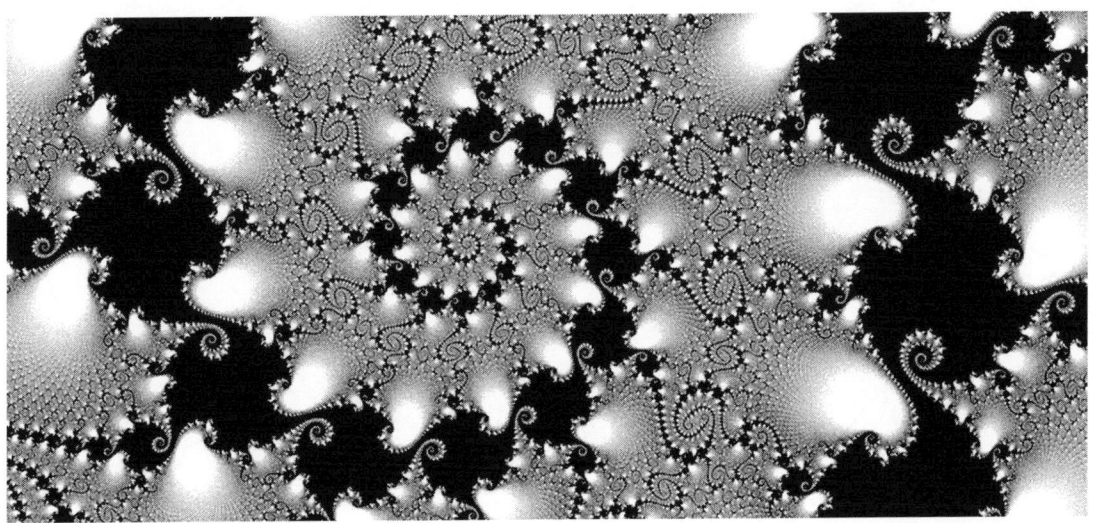

In Chapter 4, I quoted the mathematician (and author of numerous popular maths books) Ian Stewart as having written:

> *"Mathematics is full of surprises. Who would have imagined that something as straightforward as the natural numbers (1, 2, 3, 4, ...) could give birth to anything so baffling as the prime numbers?"* [20]

It's worth thinking about this for a moment. We all know what Ian Stewart means by *"Mathematics is full of surprises"* – something like "mathematics continues to surprise us". However, it's slightly misleading to express this in the way that he does. Surprise is a familiar reaction which occurs when something enters our consciousness and differs from what we expected. To a mathematically advanced civilisation which was millennia ahead of ours in its understanding, nothing that surprises us would be surprising. So it would be fair to say that the surprises are not *in mathematics*, they're *in our minds*.

There's a widespread tendency to think of the number system as "our" number system, and it's almost implied in remarks like Stewart's that it ought to "behave itself" but sometimes it doesn't (in which case we're surprised).

I find this to be an extraordinary situation. The framework for understanding reality which is propagated by Western civilisation is founded on counting and mathematics, and yet somehow *the mathematics continues to surprise us*. The basic concept of prime numbers is readily understood by ten-year-olds, yet Stewart has (quite understandably) described them as "baffling".

If something surprises us, then our expectations or assumptions about it must have been wrong. So perhaps in some deep and subtle sense *we're thinking about the number system in the wrong way* – we've got the wrong idea about it. After all, the better you get to know a person (or other animal), the less likely they are to surprise you. Likewise, the better you get to know the weather patterns in a particular location, the less likely they are to surprise you.

If we had a clear and thoroughly comprehensive understanding of the number system, then it wouldn't be able to surprise us. But it does, continually. Ian Stewart admits this when he writes "*Mathematics is full of surprises.*"

The Journey

Suppose a group of archaeologists were to dig up a spark plug from deep beneath an ancient, seemingly undisturbed site. They'd no doubt be very surprised, wondering "How did this get here? What's going on?" Comparable surprises occasionally happen in the sciences, but we're able accept this because although our scientific knowledge continues to expand, we're well aware that it's still limited.

If our scientific knowledge were *unlimited* and we knew about everything in the entirety of space and time in infinite detail, then we could *never* be surprised. Because we *don't* know everything, we speculate, theorise, hypothesise, guess, wishfully think and extrapolate in order to create a provisional "map of reality". This is reasonable enough behaviour. But it's important to keep in mind that when the map turns out to be wrong, that's *our* fault, not the universe's.

If you expect to find something and you don't find it, you're surprised. If you expect something to be true and it turns out to be false, you're surprised. But if you had no map, that is, no expectations whatsoever about those aspects of the universe you didn't already know about, then nothing could ever surprise you. So surprise is dependent on expectation.

MORE SPECULATION

All of this surprise relating to the prime numbers suggests to me that at some deep level of mind where number concepts weave into our consciousness (and our perception of time) something is wrongly configured, oriented or aligned. As with my use of the word "disease" in Chapter 38, "wrongly" isn't meant in a moralistic sense. It's more like a bone in your body being wrongly aligned as the result of an injury. Such an alignment is "wrong" because it can cause you discomfort and might lead to further damage to your body. There's no moral issue involved here, just a recognition that the body is a coherent whole and that for each part, there's an appropriate relationship with the other parts.

In analogy with this anatomical situation, I'm suggesting that among Westernised humanity, our "mental alignment" regarding the number system has somehow gone wrong, that there's a fundamental flaw in how we've been relating to number. And as the number system acts as a kind of bridge between mental and material reality, this then suggests that there may be some serious problems with the way we conceptualise and thereby relate to the world. Once more, I should emphasise that if there's any truth in this analysis of the current state of the Western collective psyche, then it's almost certainly going to be a partial truth.

You'll notice that I'm not making a moral judgement or seeking to blame anyone. But considering the place of the number system at the heart of our current worldview, that we appear to be "coming at it from the wrong angle" is potentially quite

The Journey

problematic and we might expect to see certain sociological and mass psychological symptoms of this "misalignment". So, what might the remedy be? I've suggested that certain developments in the mathematical sciences might indicate that humanity is approaching a turning point in its relationship with the number system and that some kind of beneficial change could eventually follow. James Sylvester's remark, analysed at some length in Chapter 34, suggests that a change in our relationship with the number system might require a fundamental change in our perception of *time*. How that might come about is another question.

I introduced the word "quantocentrism" in order to describe the tendency to relate to the world almost exclusively in terms of quantification and measurement. This I see as a symptom of the "misalignment" just discussed, so perhaps we need to reassess the qualitative approach to number. As mentioned in Chapter 1, in Westernised cultures this primarily manifests as numerology. Mathematicians have very little time for numerology – unlike their subject, it all seems so arbitrary, with endless contradictory interpretations in circulation and no means to confirm or refute any of the claimed properties or qualities of various counting numbers. Despite trying to keep an open mind regarding the ultimate nature of number, for similar reasons I can't endorse any particular system of numerology. Although the numerological systems of the Jewish *Qabalah* and the Assyro-Babylonian *gematria* can make fascinating study, insights gained from these will always be unverifiable outside the subjective realm of personal experience and interpretation.

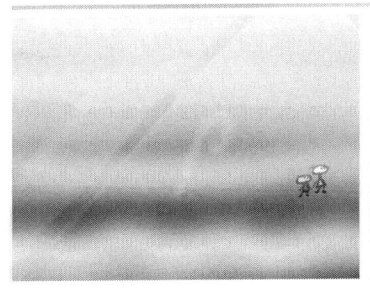

On the other hand, although it obviously treats counting numbers as quantities, classical number theory *does* deal with certain qualities – for example, whether or not a counting number is a prime, a square (a counting number multiplied by itself), a sum of squares, or the sum of all of its divisors. In this sense, some aspects of number theory could almost be seen as "verifiable numerology". But this involves no reference to the contents or events of the physical world, whereas familiar systems of numerology generally do.

Pre-Westernised cultures often related to number in a qualitative way. Although it would be pointless for us to try to adopt this approach (stuck as we are inside our quantitative mental framework), we might do well to humble ourselves and seek a deeper understanding of these cultures' relationships with the number system. There's a field of study called *ethnomathematics*, but this is quite marginal and much more closely aligned with anthropology than with mathematics, so most mathematicians are barely aware of it and not particularly interested. The dominant view within the mathematical establishment is that although anthropologists might learn something about the cultures they're studying through an examination of their number-related concepts, these cultures have nothing to teach *us* about the nature of the number system. It's hard to see how they could in any concrete way, but I feel that we should at least leave open the possibility that they *might*, in a way that no one has yet anticipated. After all, centuries of intensive, detailed research in the mathematical sciences have led Western culture into an unexpected situation where "our" number system is no longer behaving like the entity we once believed it to be.

The Journey

Compared to all others, Western culture has accumulated a huge amount of knowledge about the workings of the number system, but it seems that it's still missing something of fundamental importance. I sometimes wonder whether pre-Westernised cultures have been seeing something about it which has drifted from our view as we've sharpened our focus. It might be something which mathematics as we know it is never going to be able to deduce. It may even be somehow related to that great mystery lurking behind the zeros of the Riemann zeta function.

So perhaps an "appropriate relationship" with the number system might be established through the mass cultivation of a qualitative appreciation of number to complement our quantitative approach. But it's hard to imagine how that might occur, apart from the unlikely scenario wherein the study of classical number theory suddenly becomes an enormously popular activity akin to Rubik's Cubes in the 1980s or the early 21st century Sudoku craze.

chapter 40
number and psyche

For years, clinical psychologists have been reporting an unusual phenomenon occurring in people affected by various types of autism. In some cases, they seem to have an extraordinary intuitive affinity with the number system, and this often includes a fascination with prime numbers. This came to widespread public attention with the publication of Mark Haddon's 2003 novel *The Curious Incident of the Dog in the Night-time*, written from the point of view of a prime-number-obsessed teenager seemingly[1] with Asperger syndrome.

Neuroscientist Oliver Sacks' 1985 book *The Man Who Mistook His Wife for a Hat* contains a chapter in which he describes his contact in the mid-60s with a pair of autistic twins who possessed the ability to recognise prime numbers of up to ten digits. The twins, he claims, would spend time together, spontaneously exchanging large prime numbers "*like two connoisseurs wine-tasting, sharing rare tastes, rare appreciations.*"[2] Eventually realising what they were doing, he obtained a book of large primes and was able to interact with them, much to their delight. Sacks' description suggests that the twins had (as some other autistic people seem to have) a kind of direct access to the "landscape" of number, rather than just an ability to carry out rapid mental calculations.

Enrico Bombieri, commenting on this story, wrote

"*It is hard for me to hear this story without feeling awe and astonishment at the workings of the brain. But I wonder: Do my nonmathematical friends have the same response? Do they have any inkling how bizarre, how prodigious and even otherworldly was the singular talent the twins so naturally enjoyed? Are they aware that mathematicians have been struggling for centuries to come up with a way to do what John and Michael did spontaneously: to generate and recognize prime numbers? Or can most people do little more than shrug and perhaps secretly imagine that a real mathematician would find what the twins did no more taxing or worthy of attention than performing long division in one's head?"* [3]

The implications of the story are tremendously exciting, for it suggests that the human mind, at least in some special cases, can interface with the number system in a way which is completely alien to conventional mathematics. Unfortunately (for those of us who find this possibility exciting), the accuracy of Sacks' claims was called into question by Makoto Yamaguchi of Waseda University's Department of Educational Psychology in an article published in the *Journal of Autism and Developmental Disorders*. The issue was whether or not Sacks, as claimed in his book, could have had access to a book of ten-digit primes in 1966. Yamaguchi explains:

"I contacted him recently and asked about it. According to him, not only the book but also other resources are lost now. He admitted that the book may have included only smaller numbers (e.g., up to 8-digit).

Although I do not doubt that the twins had exceptional number abilities and that the report was a basically true story, one should not literally believe exact details of that report. Also notice the existence of skeptical views. (For instance, Dehaene, 2001, claimed that such reported abilities might have been 'hagiography')." [4]

The Journey

Regardless of the exact truth of the story, there are other cases involving autistic abilities to interface with the number system in inexplicable ways. One example is the autistic savant Daniel Tammet. He's able to instantly recognise large prime numbers, describing his experience of them in his 2006 memoir *Born on a Blue Day* as feeling "*smooth and round…similar to pebbles*". They are "*distinct from composite numbers …that are grittier and less distinctive.*" He elaborates:

> "*Whenever I identify a number as prime, I get a rush of feeling in my head (in the front center) which is hard to put into words. It's a special feeling, like the sudden sensation of pins and needles…Some nights, when I'm having difficulty falling asleep, I imagine myself walking around my numerical landscapes. Then I feel safe and happy. I never feel lost, because the prime number shapes act as signposts.*"[5]

Tammet is a particularly special case because he has both this unusual perception of numerical reality *and* a highly developed ability to articulate his perceptual experiences (he has savant syndrome[6] as well as Asperger syndrome, a very rare combination). There may be many other autistic individuals with a similar type of experience of the number system, but without the ability to communicate this to others.

The existence of such exceptional individuals adds to the overall feeling that there's something important missing from the current academic understanding of the number system's true nature. By interacting with some of them in a spirit of open-minded exploration, the mathematical community could perhaps learn something of real importance.

Although there's no reason to believe that he had any form of autism, the legendary Indian mathematician Srinivasa Ramanujan (1887–1920) appears to have had some kind of mysterious access to the number system which differed profoundly from that of ordinary academic mathematicians. As mentioned in Chapter 22, he had a reputation for "intuitively" producing extraordinarily difficult results in number theory, attributing his abilities to the inspiration of his Hindu family goddess Namagiri [7]. He's reported to have said that *"an equation has no meaning for me unless it expresses a thought of God."* [8]

The topic of religion has already made a couple of appearances in this trilogy. In Chapter 22, I suggested that the Riemann zeta function could, in an a parallel world, easily have become an object of religious devotion. Although I was very careful to avoid *advocating* this kind of behaviour, I also observed that there would be something fitting about it, bearing in mind my suggestion in Chapter 1 that number (via economics and science) represents the true religion of the modern West.

I detect a certain religious awe in some of the language used by people writing about issues related to prime numbers. Words and phrases like *secret source, profundity, profound mystery, great mystery, arcane music, secret harmony, Nature's gift, inexplicable secrets of creation, heart, soul, cosmos, divine, Holy Grail, Lucifer, Devil* and *God* have all appeared in this context.

I've also witnessed this kind of awe when explaining some of these ideas to friends and

The Journey

people I meet. Two examples come to mind. One friend of a friend, a bright young web-designer, when shown an image of the nontrivial zeta zeros (having first been given a quick tutorial about the Riemann zeta function, *etc.*), described the experience as like looking at "the face of God". Another (with more mathematical experience and no particular religious beliefs), having received a thorough explanation of the main ideas involved in this trilogy, reverted to Old Testament imagery, comparing the experience to "sneaking into the Holy of Holies and having a look inside the Ark of the Covenant"!

In the process of maintaining a website related to these issues for over a decade[9], I've become aware of a body of quasi-religious or mystical writings involving prime numbers. A lot of this I find quite confused (and therefore confusing), as the authors clearly haven't understood some of the basic mathematical ideas involved. Even when I don't take it seriously from a mathematical point of view, though, I find that some of this material can be instructive, illustrating certain extreme cases and aberrations in how individual humans can "relate to the number system". A lot of it also seems infused with a kind of longing for *a different kind of relationship* with the number system. I've witnessed this kind of longing consistently in various New Age circles, where numerology, "sacred geometry", fractals, spirals and geometrical crystal forms all have a strong following, in conjunction with an equally strong distrust of economics- and science-based thinking.

As I explained in Chapter 5, there are people looking for hidden meanings and messages in the primes, trying to "crack the code" of the universe. There are people trying to relate the primes (never very convincingly) to various fundamental physical constants[10], the structure of DNA, Ayurveda[11], the Bible or the Qur'an. These are all things which the authors, from their various perspectives, see as fundamental to the structure of reality, or "God-given". My suspicion is that, on learning of the fundamental nature of the primes, they feel the need to make these links in order to reinforce their own particular belief system.

I've seen a page on a Christian website dedicated to a number theoretical curiosity called the *Ulam spiral phenomenon*. This was discovered by the nuclear physicist

Stanislaw Ulam as a result of some doodling during a boring lecture in 1963. He wrote the positive integers out in a kind of "square spiral" and then circled the primes...

 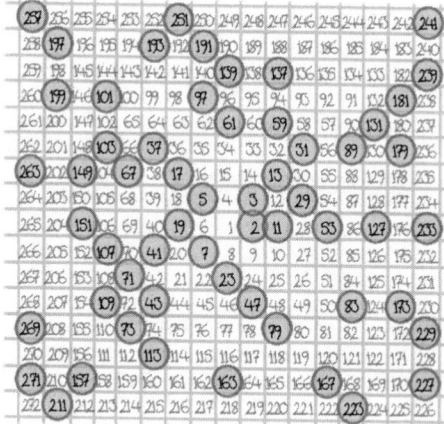

...finding, much to his surprise, that there were pieces of diagonal line showing up. Expanding the spiral and zooming out, you see this:

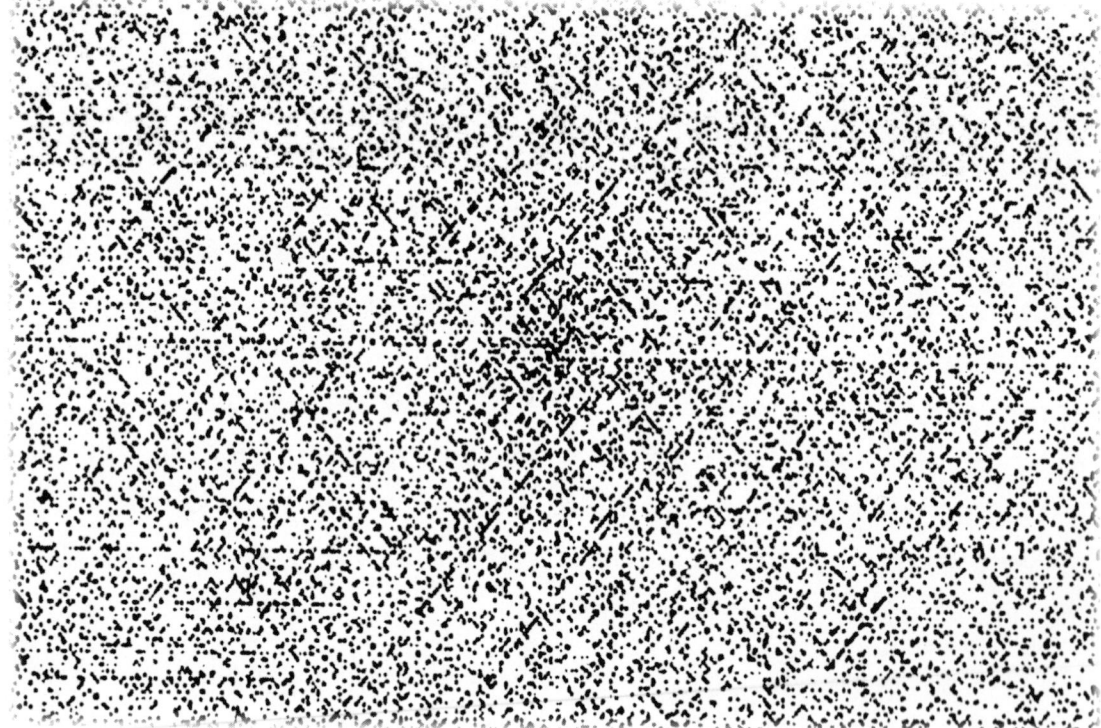

Although this phenomenon hasn't yet been fully accounted for, it's not seen by mathematicians as anything of any great importance. Expressed in number theoretical terms (rather than as a geometric pattern), the phenomenon in question is that for many integer constants b and c, numbers of the form $4n^2 + bn + c$, for $n = 1, 2, 3,...$ seem to have an unusually high probability of being prime (compared to other numbers of a similar magnitude). It's the kind of thing that may well be proved eventually, so what we're considering here is just an unproved number theory conjecture that happens to be expressible as a visual pattern of sorts.

The Christian website, however, was presenting the Ulam spiral phenomenon as nothing less than *evidence that God exists*[12]. If the authors made the effort to learn about the Riemann zeta function, they'd almost certainly interpret what they learned as even more convincing evidence that God exists. But whether or not God (as understood by them) exists isn't the point here. What interests me is a particular *feeling* linked to our current relationship with the number system, which is manifesting in these religious authors as a wish to connect the mysteries of the (eternal, ineffable) number system to their idea of (eternal, ineffable) God.

There are also feeble "theological" arguments which have been presented as proofs of the Riemann Hypothesis, as well as tremendously confused books like Peter Plichta's *God's Secret Formula: The Deciphering of the Riddle of the Universe and the Prime Number Code*[13] and Adri de Groot's self-published *Number Theory in Light of Unification Thought*[14] (which claims to verify the teachings of the Reverend Sun Myung Moon).

The Journey

This fragment of questionable science journalism…

> "…there is a growing belief among scientists and mathematicians, a growing faith of religious proportions, that somehow prime numbers are the key to solving the greatest philosophical mystery of all: why there is something instead of nothing."[15]

…is quite misleading. I've never come across *any* scientists or mathematicians who admit to holding this view. There certainly is no "*growing faith of religious proportions*" in 2013 (this was written in 2003). Wherever the author got this misguided notion, though, the fact that a science writer believed it to be true tells us something.

My overall impression of this situation is that because the Western "map of reality" has been built on top of the number system, when the number system's mysterious underlying structures become known to a religiously-oriented person, these can take on religious significance. But worldwide, we find religions in which certain numbers have a special status (the Christian Trinity and the twelve apostles, the "eightfold path" in Buddhism, among many others). Number has religious connotations in many cultures, so it could be plausibly argued that those aspects of the mind which interface with the number system are naturally linked to those aspects that deal with religious belief, experience and ritual.

REVEALING LANGUAGE

Here are some quotations from various mathematicians which contain unusually poetic, "ecstatic" and/or mystical language (I've added the boldface emphases):

"…there is no apparent reason why one number is prime and another not. To the contrary, upon looking at these numbers one has the feeling of being in the presence of one of the **inexplicable secrets of creation**." (Don Zagier)[16]

"The **mystery** that clings to numbers, the **magic** of numbers, may spring from this very fact, that the intellect, in the form of the number series, creates an infinite manifold of well-distinguished individuals. Even we enlightened scientists can still feel it, e.g. in the impenetrable law of the distribution of prime numbers." (Herman Weyl)[17]

"To me, that the distribution of prime numbers can be so accurately represented in a harmonic analysis is **absolutely amazing and incredibly beautiful**. It tells of an **arcane music and a secret harmony** composed by the prime numbers." (Enrico Bombieri)[18]

"Riemann had found a passageway from the familiar world of numbers into a mathematics which would have seemed utterly alien to the Greeks who had studied prime numbers two thousand years before. He had innocently mixed imaginary numbers with his zeta function and **discovered, like some mathematical alchemist, the mathematical treasure** emerging from this admixture of elements that generations had been searching for."

"Until [the RH is proved], we shall listen **enthralled** by this unpredictable mathematical music, unable to master its twists and turns. The primes have been a constant companion in our exploration of the mathematical world yet they remain the most **enigmatic** of all numbers. Despite the best efforts of the greatest mathematical minds to explain the modulation and transformation of this **mystical music**, the primes remain an unanswered **riddle**." (Marcus du Sautoy)[19]

Marcus du Sautoy's 2003 book *The Music of the Primes* is full of poetic language like this. Some of the words and phrases used are *music* (repeatedly), *sense of wonder, timeless, Nature's gift, secret source, inner harmony, elixir, stunning, yearning, misty waters, vast ocean, vast expanse, awesome vista, looming out of the mist, unleash the full force, radically new vistas, hidden harmonies, poetry, jewels, crown, riddle, cryptic, most enigmatic, mystical ley line* and *diabolical malice*.

Elsewhere in popular mathematics literature, we find passages concerning the primes, zeta function and RH involving words like *mystery, mysterious* and *secrets* (numerous times), *strange, stunning, astonishing, baffling, bafflement, devilment, bedevilled, strange fascination, yearning, mysterious attraction, incredible, exalted, majestic, fantastic, miraculous, amazed, absolutely amazing, awed, impenetrable, impenetrability, unveil, blazed ... fearlessly, glory, ancient, quest, profundity, profound mystery, great mystery, gemstone, heart, soul, cosmos, abysses, divine, Holy Grail, Lucifer, Devil* and *God*.

This kind of language is aligned with a certain quality of feeling not generally associated with mathematics. As I expressed earlier with regard to the quantum chaological connections, *it's hard not to wonder what it is we're ultimately dealing with here*. Mathematicians aren't generally known for expressing this kind of feeling (one which we're more likely to associate with poets and mystics), apart from occasional references to the flashes of inspiration which seem to come from somewhere outside themselves. Number theorists don't seem any more inclined to express such feelings

The Journey

than other types of mathematicians. How, you might wonder, can they be immersed in this stuff, with the deeply mysterious qualities we've been looking at, and not be in a permanent state of awe?

A partial answer is that mathematical research is an extremely demanding and time-consuming activity, so too much contemplation about "what it all means" is generally an obstacle to progress. Another is that people can assimilate and integrate surprising facts until they start to seem normal. Surprise, by its nature, always fades away before too long. This is necessary in order for us to function – to be in a permanent state of surprise would be seriously debilitating! You're taken by surprise when you find out that something isn't the way you thought it would be. Once you've absorbed the fact that it really *is* like that, as surprising as it might initially have seemed, the surprise dissipates – as you now *know* that it's like that.

Each layer of surprise is gradually absorbed and forgotten. Perhaps because of this, no one (it seems) has noticed the overall phenomenon of people being surprised by this prime-number-related material. Despite having combed through all the relevant literature, I've never come across *any* discussion of this phenomenon.

chapter 41
number and the other

Consider these words, extracted from the same body of popular literature on prime numbers and the zeta function which I mentioned at the end of the last chapter:

Mysterious, strange, stunning, astonishing, baffling, exasperating, perplexing, stultifying, cruelly compelling, fascinating, obsession, mysterious attraction, most beautiful, incredibly beautiful, breathtakingly beautiful, immense beauty, beautiful harmonies, elegant, elegance, gorgeous, glamorous, impenetrable, tantalized, tantalizing, tantalizingly vulnerable, unveil.

A few years ago, I discussed this use of language with a couple of thinkers who are versed in both Jungian/transpersonal[1] psychology and higher mathematics, Barry Jeromson and Robin Robertson. Jeromson's doctoral thesis was entitled "Jung and mathematics in dialogue: A critical study", while Robertson has been an editor of *Psychological Perspectives* (the journal of the C.G. Jung Institute of Los Angeles) since 1986 and has written numerous books on Jungian thought. Both have offered some interesting insights, so I'll include some excerpts from our email discussions. In reading these, it would be helpful to have a basic grasp of such Jungian concepts as *archetypes*, *anima* and *animus*, the *shadow*, *compensation*, *projection*, the four-way scheme of the *thinking*, *feeling*, *sensation* and *intuition* functions, *etc*. If you're unfamiliar with these, some explanatory notes can be found in Appendix 16.

In an initial response to the list of words and phrases I list above, Jeromson commented:

Many of these adjectives are also used to describe a beautiful woman, especially one who is aloof and remote. Are the mathematicians projecting their animas onto the mathematics? In the words of the post-Jungian writer James Hillman, the mathematics 'anima-tes' their imaginations. Incidentally the use of the term 'elegant' to describe a proof lends weight to this argument.

Some time later, I discovered that André Weil, in his review of Emil Artin's *Collected Papers*, wrote[2]:

> "*Perhaps the best part of this career may be described as a love affair with the zeta function.*"

In a similar vein, Jean-François Burnol characterises the Riemann Hypothesis as a "distant" female in the following excerpt from his 2008 paper "Fourier and zeta(s)"[3]:

> "*What is more, Theorem 6.30 has encouraged us into trying to encompass in our speculations the GUE hypothesis, and more daring and distant yet, the Riemann Hypothesis Herself.*"

In Karl Sabbagh's book *Dr. Riemann's Zeros*, Alain Connes is quoted, again invoking the feminine[4]:

> "*I believe I have found a very nice framework* [to prove the Riemann Hypothesis] *but this framework is still awaiting the main actor. So there is the stage – it is perfectly well arranged and so on – but we are still expecting the heroine to come and complete it.*"

Notice the use of "heroine" rather than "hero".

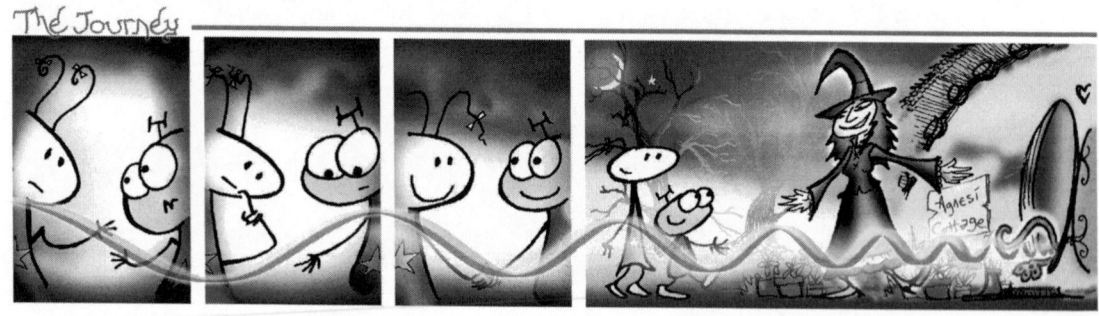

The early 20th century number theorist John Littlewood, according to Béla Bollobás' foreword to a new edition of *Littlewood's Miscellany*, became *"infatuated"* with the problem of the Riemann Hypothesis and *"was not alone in that"* [5].

A *Village Voice* review [6] of the popular books by Sabbagh, du Sautoy and Derbyshire observed that *"In discussing the primes, mathematicians often use the vocabulary of first love"*:

This is from John Derbyshire's 2003 book *Prime Obsession*:

"*In* [his 1859 paper], *Riemann made an incidental remark – a guess, a hypothesis. What he tossed out to the assembled mathematicians that day has proven to be almost cruelly compelling to countless scholars in the ensuing years…*

…it is that incidental remark – the Riemann Hypothesis – that is the truly astonishing legacy of his 1859 paper. Because Riemann was able to see beyond the pattern of the primes to discern traces of something mysterious and mathematically elegant at work – subtle variations in the distribution of those prime numbers…

It has become clear that the Riemann Hypothesis, whose resolution seems to hang tantalizingly just beyond our grasp holds the key to a variety of scientific and mathematical investigations… Hunting down the solution to the Riemann Hypothesis has become an obsession for many…"[7]

Note *"cruelly compelling"*, *"mysterious and…elegant"*, *"seems to hang tantalizingly just beyond our grasp"* and *"obsession"*. The choice of language speaks for itself.

Here are some more relevant passages:

"*Let us now pursue an apparently tangential path. We wish to consider one of the most fascinating and glamorous functions of analysis, the Riemann zeta function...*" (Richard Bellman)[8]

"*...the greatest mathematicians of all ages have found in* [the theory of prime numbers] *a mysterious attraction impossible to resist.*" (Godfrey H. Hardy)[9]

"*Hardy grew to love the* [Riemann Hypothesis]. *He and Littlewood wrote at least ten papers on the zeta-function.*" (J. Brian Conrey)[10]

"*...Gauss liked to call* [number theory] *'the Queen of Mathematics'. For Gauss, the jewels in the crown were the primes, numbers which had fascinated and teased generations of mathematicians.*" (Marcus du Sautoy)[11]

"*Even Alan Turing...was seduced by the fascination of the Riemann Hypothesis. In the midst of laying the theoretical foundations of what were to become digital computers, Turing designed a machine to calculate zeros of the Riemann zeta function.*" (Karl Sabbagh)[12]

The words "tantalise", "teased", "seduced", *etc.* appear repeatedly in this context, which is very much compatible with Robertson's suggestion of an "anima projection" at work:

"*A particularly tantalizing aspect of the chaotic scattering process is that it may connect the mysteries of quantum chaos with the mysteries of number theory.*" (Martin C. Gutzwiller)[13]

The Journey

"We have presented several tantalizing connections between xp and ζ(s). However it is clear that more is required to transform our hints and guesses into an unambiguous and satisfactory construction of the Riemann operator."

"Our purpose is to report on the development of an analogy, in which three areas of mathematics and physics, usually regarded as separate, are intimately connected. The analogy is tentative and tantalizing, but nevertheless fruitful. The three areas are eigenvalue asymptotics in wave (and particularly quantum) physics, dynamical chaos, and prime number theory." (Michael Berry and Jonathan Keating) [14]

"…in one of those unexpected connections that make theoretical physics so delightful, the quantum chaology of spectra turns out to be deeply connected to the arithmetic of prime numbers, through the celebrated zeros of the Riemann zeta function: the zeros mimic quantum energy levels of a classically chaotic system. The connection is not only deep but also tantalizing, since its basis is still obscure – though it has been fruitful both for mathematics and physics." (Michael Berry) [15]

"The primes have tantalized mathematicians since the Greeks, because they appear to be somewhat randomly distributed but not completely so." (Timothy Gowers) [16]

Here's a dictionary definition of "tantalise": *To tease or torment by or as if by exposing to view but keeping out of reach something that is much desired.* [17]

Could the tendency to use words like this be evidence of the unconscious mind personifying the zeta function and related mathematical structures as teasing,

tormenting entities, keeping something desirable just in view, but out of reach? Here are a couple more:

> "*Berry and his collaborator Jon Keating* [showed] *how techniques in number theory can be applied to problems in quantum chaos and vice versa. In itself such a connection is very tantalising. Although sometimes described as the Queen of mathematics, number theory is often thought of as pretty useless, so this deep connection with physics is quite astonishing.*"

> "*Berry is also convinced that there must be a particular chaotic system which when quantised would have energy levels that exactly duplicate the Riemann* [zeros]. *'Finding this system could be the discovery of the century,' he says. It would become a model system for describing chaotic systems in the same way that the simple harmonic oscillator is used as a model for all kinds of complicated oscillators. It could play a fundamental role in describing all kinds of chaos. The search for this model system could be the holy grail of chaos…Berry believes the system is likely to be rather simple, and expects it to lead to totally new physics. It is a tantalising thought.*" (Julian Brown)[18]

When shown the following remark from the great Paul Erdős…

> "*It will be another million years, at least, before we understand the primes.*"[19]

…my friend Marcus Clark (who originally introduced me to Jungian thought) said this reminded him of an exasperated man talking about his wife after an unresolved argument.

Incidentally, Erdős was never married nor (as far as we know) ever expressed any interest in romantic relationships. I should clarify at this point that I'm certainly not in any way attempting to draw conclusions about the individual psychologies of anyone quoted in this chapter. I see the issue here as a *collective* phenomenon, and those quoted are simply acting (unintentionally) as conduits for something going on in the collective unconscious. The invoking of "the feminine" is significant not for what it suggests about the individual mathematician's psyche, but as an indication of a collective encounter with what continental philosophy would call "the Other" [20], that is, something fundamentally alien to the Western psyche as it's currently configured. (This concept has been applied to such diverse topics as purported encounters with UFOs, fairies and angels, the use of racist stereotypes in wartime propaganda and the analysis of dreams.) In the Jungian analysis of male-female relationships, the individual experience of the other person is a particular manifestation of a wider psychic phenomenon of encounter with the Other.

Although the preceding paragraph might seem clumsy to anyone properly versed in Jungian thought and/or continental philosophy, I hope I've conveyed the central idea – that is, that mathematicians have reached a point in their collective exploration of the mathematical landscape where they are now encountering something fundamentally *other*. And some mathematicians have felt compelled to comment on this. As mathematics constitutes the core of Western science and, in fact, of the entire "scientistic" civilisation which has almost entirely excluded "the feminine"

from its considerations, these individuals have instinctively drawn on the language of the feminine, romantic attraction, *etc.* in order to convey this sense of "otherness".

On a related note, in his *Jerusalem: City of Mirrors*, Amos Elon observes the recurrence of feminine metaphors associated with Jerusalem through the ages and concludes (I quoted this in Chapter 22, but it's worth repeating):

> "*Language was playing odd tricks on the patriarchal East. Public statements are often rooted in private dreams. When men are mystified, they often resort to the feminine gender.*" [21]

As of 2013, the remarkable things I've attempted to explain throughout this trilogy are still considered to be parts of marginal, specialised topics within the mathematical sciences. But through the inevitable spreading of awareness about this "Otherness" of the number system which I've attempted to convey, the (male-dominated) scientific world could be on the edge of a collective psychic "shock" which will be somehow analogous to the (personal) psychic upheaval that can occur the first time someone falls into a serious romantic relationship. Could it be that mathematics is approaching a transition of a similar magnitude to that which occurred in physics with the advent of quantum mechanics in the early 20th century?

In Chapter 37, I put forward the possibility that the distribution of primes could

The Journey

have some significance in terms of Jung's concept of *archetypes*. Continuing to explore in this direction, we come to the idea that the pursuit of the elusive proof of the Riemann Hypothesis is something like a "grail quest" (as in the Arthurian literature, the multi-levelled symbolism of which has been of great interest to Jungian/transpersonal psychologists). The Riemann Hypothesis has been described as a "Holy Grail" by several commentators. For example, writing on Hans Rademacher's failed 1945 disproof attempt at the RH, Karl Sabbagh remarks:

> "*It was not the first occasion, and would not be the last, on which a distinguished mathematician believed he had within his grasp the Holy Grail of number theory.*"[22]

Robin Robertson: That's fascinating. I like the idea of a mathematical grail quest. That really brings in the archetype of the feminine you mention... I assume you know the Jungian stuff on the grail quest as a search for the missing feminine. In essence, Parsifal had to find the feminine inside himself in order to find the projected feminine. And, of course, it was too early in time for him to fully succeed. Put that bluntly, it sounds trivial, but there's a lot in it.

MW: Barry Jeromson pointed out that a lot of the emotional adjectives used by mathematicians in regard to the distribution of primes: "strange", "mysterious", "astonishing", "breath-taking", etc. suggest an anima projection. A friend of mine has taken this idea a bit further, in terms of a compensation from the

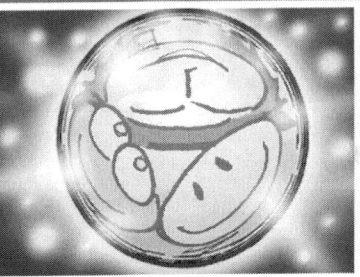

shadow, due to an overemphasis of mathematicians on the thinking function. As he put it, mathematics (and with it science) may be about to be confronted with its feminine side.

RR: Your friend sounds right on the money to me. I've been heavily involved with a chaos theory society for a number of years, always as a friendly visitor, not one for whom chaos theory was central. Among those folks, especially the techies, the desire for the feminine is palpable. They're clearly thirsting for something beyond the dry, arid land of pure logic and they don't know how to find what they need.

I'd go further than Jeromson. He asked whether mathematicians are projecting their animas onto the mathematics. As I've already stated, I don't believe that this is merely about the cumulative effect of individual male mathematicians experiencing personal psychological reactions to the object of their study. In other words, this isn't just about male mathematicians neglecting or suppressing their animas and consequently experiencing a sort of compensation involving an anima projection onto the mathematics. I think we're instead seeing a *collective* compensation involving the *collective* anima being projected onto the mathematics associated with the primes and the zeta function. Global Westernised culture as a whole, increasingly concerned with technology and economics to the exclusion of all else, has been suppressing its collective anima. The compensation phenomenon applies to the whole culture, but is currently being experienced only by those explorers at the heart of number theory, the "Queen of Mathematics",

The Journey

mathematics itself being acknowledged as the "Queen of the Sciences".

The number theorists whose writings have included the poetic, ecstatic and mystical language reproduced in the previous chapter belong, I've already argued, to the inner priesthood of this "scientistic" civilisation. I'm speculating that they may thus be among the first to be encountering some collective psychological reaction brought about by a mass psychic imbalance associated with "quantocentrism" – that is, a collective obsession with quantity.

This is from an email I received from Alex Abercrombie, a retired mathematician also familiar with Jungian thought:

You've undoubtedly tumbled to something with your idea of a collective anima projection. My belief is that this has to do with the anima being part of the interface between conscious and unconscious. The key moment in mathematics is when you suddenly find that an idea has a life of its own – you define something with a certain purpose in mind and then you discover other entirely natural properties which weren't intended in the definition. Of course a prime example (excuse the pun!) of this is the primes, which are boring enough if you just consider them as generators of the multiplicative semigroup on $\mathbf{Z}$. But throw in the order on $\mathbf{Z}$ – not to mention the additive structure – and suddenly you're in the world of the Prime Number Theorem, Dirichlet's theorem[23], Goldbach's conjecture and all that... So the sequence of primes

in a way embodies this business of a limited conscious idea suddenly showing signs of independent life – but at the same time showing that it still conceals at least as much as it reveals. It is a first approach of the unconscious - therefore definitely an occasion for anima projection!

It would be interesting to know whether women mathematicians are inclined to project the animus in similar situations. I rather suspect not, and if I'm right this would tend to support your idea of the anima projection being collective rather than personal.

As mentioned in Chapter 37, nearing the end of his life, Carl Jung concluded that the system of counting numbers constitutes a single archetype, the *archetype of order*. Beyond this, he thought that in modern Westernised humanity, this archetype *had become conscious*. The current Western obsession with quantity, quantification and measurement (what I've been calling quantocentrism) could perhaps be understood in these terms – the collective psychological relationship with number has become dangerously unbalanced. Jung's student Marie-Louise von Franz developed these ideas in her book *Number and Time* [24]. Von Franz died in 1998 and neither she nor Jung (as far as I know) had a chance to explore the archetypal significance of the mysterious, irregular sequence of prime numbers which reality presents to us, embedded within the sequence of counting numbers.

One problem with Jungian theory (from the point of view of quantitative science) is

The Journey

the impossibility of precisely defining the key vocabulary. So the matters discussed here are still far from clear and would probably provoke considerable debate within any group of Jungian thinkers. For example:

BJ: I have some problems with this piece of Jungian rhetoric [the 'becoming conscious' of the natural number archetype of order]. More likely, large chunks of the collective western consciousness are gripped by the number archetype. It is still unconscious and driving the ship. If the number archetype became conscious, then it would lose its power and the widespread urge to quantify everything would evaporate.

Several years after our email dialogue and having since moved away from Jungian thought, Barry Jeromson agreed to read through a draft of this chapter. This led him to put forward further reservations about this Jungian approach to number-related issues. These are outlined in Appendix 17.

NUMBER, THE FEMININE AND THE UNQUANTIFIABLE

Consider the following diagram, which attempts to (loosely[25]) represent certain aspects of human consciousness:

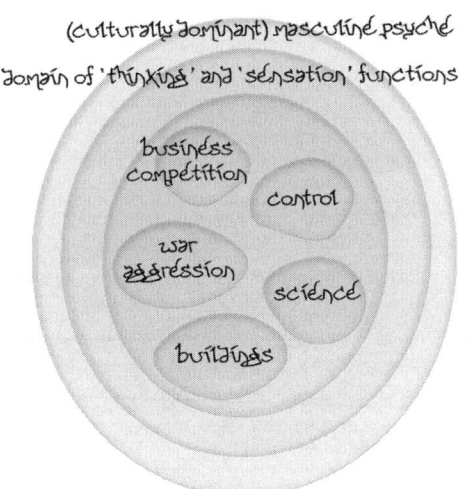
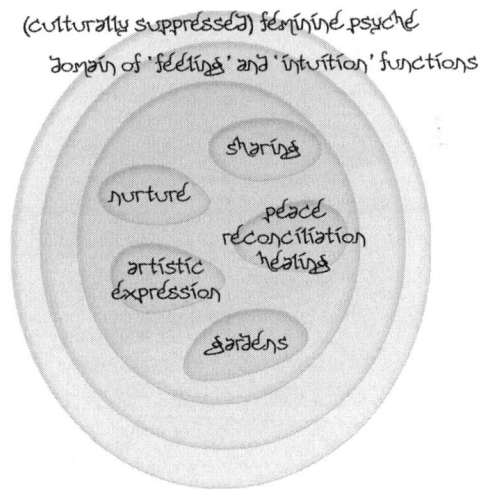

It's important to stress this masculine/feminine division is *not* a male/female division. We're talking about aspects of the psyche, all individuals having both sides as part of their psychic makeup, with differing emphases depending on gender, personality, cultural influences, sexuality, *etc*. This division may well be related to the left/right brain hemisphere division, but not in the direct way which oversimplified accounts of brain hemispheres might suggest. The basic division into *thinking-sensation* and *feeling-intuition* is based on Jung's ideas. The more detailed, somewhat arbitrary, division of human activity into {control, war/aggression, business/competition, buildings/structures, science/engineering} and {nurture, sharing/cooperation, peace/reconciliation/healing, gardens, arts} can be argued about, but it's just a loose scheme based on my own observations and experience. There's an exercise I've tried on people, where they're asked to sort cards into "masculine" and "feminine" (rather than strictly "male" and "female") piles:

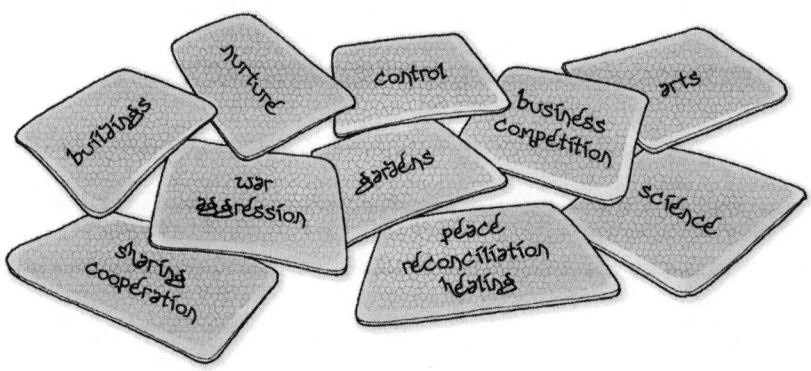

Sometimes "artistic expression" ends up on the masculine side (if asked to pair

up arts/science, though, people invariably choose feminine/masculine). Other than that the division tends to reflect the diagram shown on the previous page.

If you instead ask someone to divide the cards into piles of *what they think the world needs more of* and *what they think the world needs less of*, you'll tend to find a similar pattern of division. There will be exceptions – some insecure people might think the world needs more "control", some enthusiastic capitalists more business, a few might think it needs more buildings (builders and architects perhaps – otherwise it's hard to imagine why). At the moment, a significant proportion – mostly men – will think the world needs more science, but I would guess that this proportion is shrinking. Regardless, the "feminine" side is, on average, going to score far more points in the "what the world needs more of" game.

Without getting into the enormous question of what can be done about it, it's my opinion (and that of a huge consortium of people worldwide who get grouped together loosely as "anti-capitalist", "global justice", "green", *etc.*) that there's too much war, too much business (busyness), too many buildings, too much control, not enough nurture, not enough gardens, not enough artistic expression, sharing, healing or reconciliation. So something's fallen out of balance. It seems that we've strayed too far to one side.

Now let's try subdividing those cards according to the scheme "quantifiable or unquantifiable". Ask yourself: in which of the areas described is mathematics most readily put to use?

Mathematics can be exploited effectively in many types of "control" (economic manipulation, surveillance technologies). It feeds directly into science, and through science into warfare. Business is all about money, economics – numbers, in other words. And mathematical formulas are continually being employed in the design and construction of buildings.

Although mathematics does come into the arts in various ways, *nurture*, *sharing/ cooperation*, *peace/reconciliation/healing*, *gardens* and *artistic expression* are more likely to fall on the "unquantifiable" side of the balance.

This point shouldn't be overstretched, though – these things are never that black and white. Sharing, after all, is about trying to making quantities equal, rather than trying to make them as large as possible (business). But, generally speaking, if we were to rank the cards according to their compatibility with "quantocentric thinking", the "masculine" half, the half that "the world needs less of", would occupy the top of the list.

This isn't about women's rights or blaming men for the world's problems, it's about trying to isolate the source of the imbalance in the collective Western psyche. And it's at least an *interesting possibility* that this might have something to do with our collective psychological relationship with number.

The Journey

A possible Jungian interpretation of what's going on, as touched on in the emails I've reproduced in this chapter, is that as the result of an imbalance in the collective psyche towards thinking-sensation and away from feeling-intuition, a compensatory mechanism has been activated. This has taken the form of a collective "anima projection", evident in certain parts of the (male-dominated) mathematical community. The Western psyche has, in its journey into the heart of the number system, encountered something *other*, something it has no categories for, no way of accounting for (hence the repeated expressions of surprise), but *also* something that it longs for, yearns for, desires and pursues, despite its distant, aloof, tantalizing, unattainable qualities.

Jungian theorists regularly deal with the phenomenon of anima projection, but this is quite an unusual and unexpected subject onto which the anima is being projected: aspects of pure mathematics. Still, if we acknowledge (1) that analytic number theory represents the very pinnacle of a certain type of intellectual activity (associated with pure thought and the exclusion of all feelings and non-rational considerations), as well as (2) the fact that it deals with the core issues of the number system which is embedded in the foundations of what could be called the Western "mental operating system"[26], then perhaps it's not so surprising.

A possible Jungian argument would be that the feminine aspects of the psyche have been excluded in the process of building a quantocentric culture and allowing the thinking function to become dominant. The resulting compensation process has taken

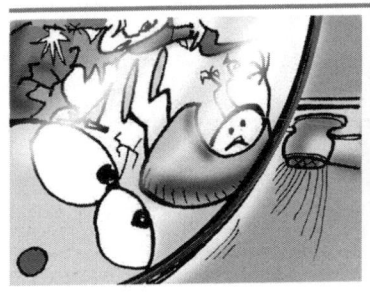 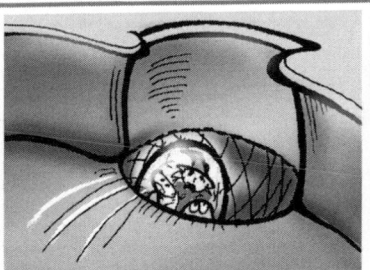 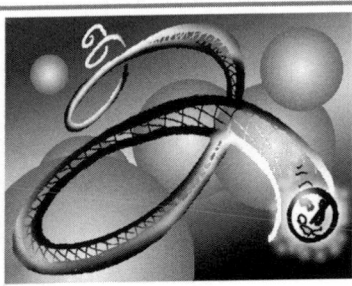

the form of a "quest" for the lost (or suppressed) feminine within the core of the thinking function's ultimate expression, that is, the scientific community. This quest hasn't yet been consciously acknowledged, but it centres on the prime numbers, the Riemann zeta function and the Riemann Hypothesis, which lie at the heart of the number system.

So, as well as the totally unexpected discoveries linking the Riemann zeta zeros to quantum chaology (and other aspects of physics), we have here a second reason for a reassessment of the number system, what it "really is", and how we relate to it. Such a reassessment cannot take place within the framework of conventional mathematical research, though, as it deals with questions of the psyche which mathematics can never address.

A reassessment of number could serve as part of a general rebalancing of the collective Western psyche, although at the moment it's very hard to imagine exactly what form this would take or how such events might unfold.

I've mentioned that Jung concluded that the sequence of counting numbers, taken as a whole, could be understood as a single archetype, the *archetype of order*, with a unique status among the archetypes. I've also pointed out that had Jung lived a few years longer, he may well have considered the role of the primes within the number system from this point of view. We'll never know what he would have had to say on the matter, but it seems to me that the sequence of prime numbers could *also* be understood as a

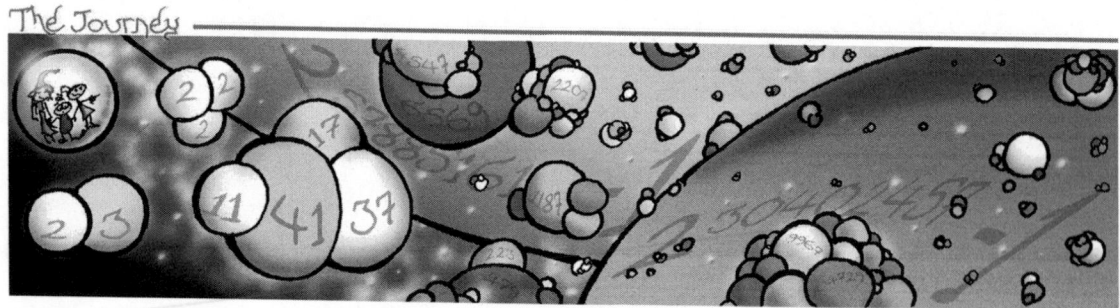

The Journey

single archetype, the "archetype of chaos" which lurks within the archetype of order. Here I'm using the word "chaos" in its older, colloquial pre-chaos-theory sense.

If there's any value in this interpretation, then we could say that modern Westernised culture has allowed the archetype of order to become dominant in its collective psyche (the "inappropriate relationship with number" which I suggested in Chapter 38). However, the natural inquisitiveness of generations of mathematicians has led to a series of revelations about the inner workings of the number system, hence the "archetype of order" has been cracked open to reveal the chaotic features at its core. This would accord with the many expressions of surprise which we've seen and also appears to correspond to the idea of the suppressed or hidden feminine being encountered as "the Other" at the core of the Western "scientistic" mind.

The idea of "chaos within order" is also resonant with qualities of many instances of *li* (the type of organic patterning discussed in Chapter 36).

THE HOLY GRAIL?

Endless surprises and aloof, unobtainable feminine Otherness… something else which has been repeatedly invoked in connection with the mysteries of the prime numbers is the *Holy Grail*. And again, this is something of considerable interest to theorists working in Jungian, archetypal and transpersonal psychology. Jungians interpret

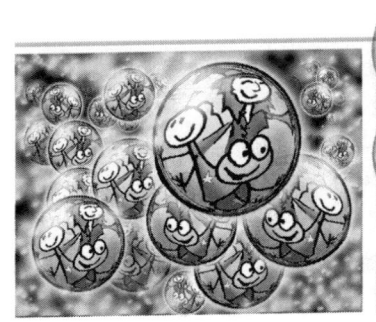
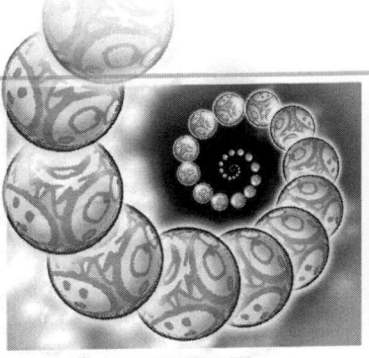
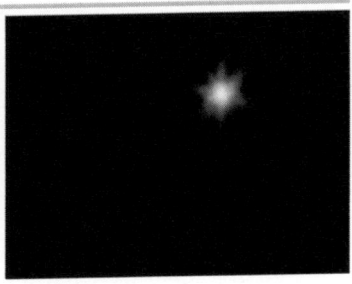

various pre-scientific systems of thought, be they mediaeval alchemy, astrology, the Tarot, the *I Ching*, the Qabalah or the Arthurian cycle of stories (including the Grail legends) as intuitive maps of the collective unconscious, produced without anyone involved necessarily having understood what they were really doing. Consequently, these systems (although Jungians generally wouldn't take them seriously in the way that some naïve modern adherents might) are seen as an indirect source of knowledge about the structure of the collective psyche.

The fact that certain correspondences can be established between these various "maps" is naturally compatible with this way of thinking. For example, the four elements of the ancient Greeks and alchemists (air, fire, earth and water) are traditionally seen to correspond to the four suits in the Tarot (swords, wands, discs and cups). These can also be put into correspondence with the Four Worlds of the Qabalistic Tree of Life [27] (*Yetzirah, Atziluth, Assiah, Beri'ah*). As well as the original interpretation associated with Christian mysticism, the Holy Grail is associated with the Ace of Cups card in the Tarot and shows up in some of the more spiritually-oriented interpretations of alchemical traditions.

For Jungians (as exemplified in Robin Robertson's email remarks earlier), the quest for the Holy Grail is linked to the quest for psychic wholeness, which, in the case of Western "scientistic" culture is a quest for the reintegration of the lost or suppressed feminine. Jungians have offered interpretations for aspects of the Arthurian Grail stories in exactly these terms.

The Journey

The Grail, a powerful symbol with a complex history, is now colloquially used to mean "the ultimately desirable entity or goal" in whatever field is being discussed. So just as a proof of the Riemann Hypothesis is described as the (current) "Holy Grail of mathematics", the Higgs boson[28] had, until very recently, been described as the "Holy Grail of particle physics". The titles of the following academic articles give some indication of this trend:

"Detecting causal relationships in distributed computations: In search of the Holy Grail" (*Distributed Computing*, 1994)

"Predicting changes in community composition and ecosystem functioning from plant traits: revisiting the Holy Grail" (*Functional Ecology*, 2002)

"T-cell responses to autoantigens in IDDM: The search for the Holy Grail" (*Diabetes*, 1996)

"Targeted Radiotherapy: Is the 'Holy Grail' in Sight?" (*Journal of Nuclear Medicine*, 2006)

"The Holy Grail of the perfect character: the cladistic treatment of morphometric data" (*Cladistics*, 1993)

"In search of the Holy Grail: Policy convergence, experimentation and economic performance" (*National Bureau of Economic Research*, 2002)

"Dipole-induced ordering in nematic liquid crystals, II: The elusive Holy Grail" (*Journal of Chemical Physics*, 2000)

A quick web-search will turn up similar references to the "Holy Grail" of, variously, genetics, quantum gravity, quantum computation, cosmology, astrophysics, statistical mechanics, biology and geology.

This kind of colloquial usage would generally be taken as an explanation for why such remarks as this...

> "Proving that the Riemann zeta function has an infinite number of solutions and that those solutions, if plotted on a graph, lie in a certain straight line that runs through the point 1/2, is considered the Holy Grail of Mathematics..." (Mike Martin[29])

...continue to show up throughout popular science and mathematics literature. However, I feel that there's more to the story. For unlike "dipole-induced ordering in nematic liquid crystals" or "the cladistic treatment of morphometric data", the Riemann Hypothesis does seem to be a worthy candidate for a "Holy Grail" in the archetypal (as opposed to the casually metaphorical or literal/historical) sense.

Having assimilated the various ideas in these last few chapters, you might be able to see what I'm hinting at here. Jungians see the Grail quest as a striving for psychic wholeness, for the rebalancing of the psyche through the reintegration of "the feminine" which has been suppressed due to the dominance of the thinking function. That would generally be in the context of an individual psyche. The *collective* suppression of the feminine can be related to the dominance of the "archetype of order" (the number system, understood quantitatively). The Riemann Hypothesis is by far the most important problem concerning the Riemann zeta function, that challenging and mysterious object at the centre of the theory of prime numbers, which is the heart of number theory (the "Queen of Mathematics", mathematics itself being the "Queen of the Sciences"). In their quest to resolve

The Journey

the RH, the mathematical community has revealed unexpected, deeply surprising aspects of the number system, and its reactions to these suggest a collective psychic encounter with "the Other".

The quest to prove the Riemann Hypothesis seems to be inevitably leading us towards a reconsideration of "what number really is". And whether or not the RH is ever proved may turn out to be less important than the unintended results on the collective psyche of this reassessment.

Epilogue

I explained in Chapter 21 how the primes and nontrivial Riemann zeta zeros exist in a kind of duality, how they're effectively "two sides of a coin". The role of the primes as the skeletal understructure or "scaffolding" of the number system should be clear by now. This then raises the following question: if the number system is so strongly linked to our framing of reality, then shouldn't the Riemann zeta function (with its zeros) be similarly linked to something of a similar depth of significance (something "dual", in some sense, to our framing of reality)?

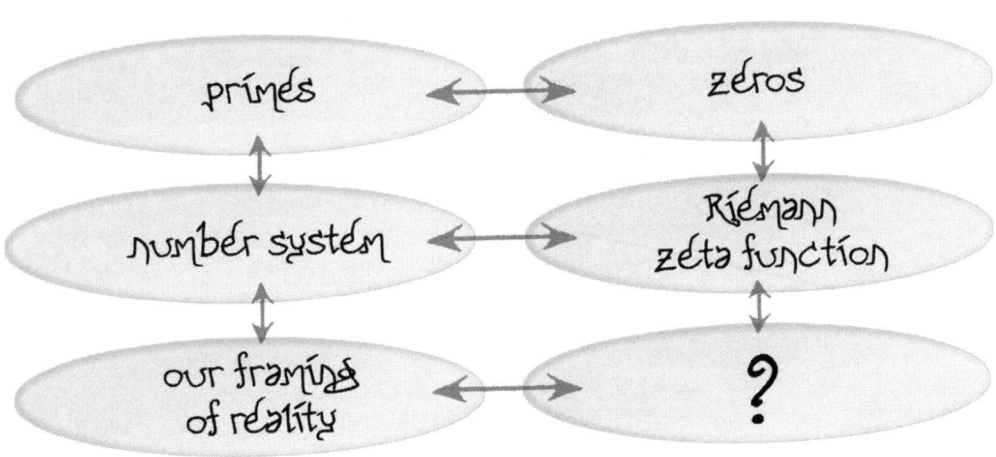

What, if anything could this be? The best clues we have were discussed in Chapter 30 when we saw how the "spectrum" of zeta zeros seems to originate with some kind of quantum chaological system and that a hypothetical "Riemann dynamics" with this spectrum could potentially even be constructible within physical reality. This might seem far too obscure or marginal to qualify for such a fundamental role in the "structure of reality" or our framing of it. But as it's still hypothetical, no one having yet observed or constructed one, it's hard to arrive at any meaningful judgment. Still, this way of thinking might help explain why the hypothetical Riemann dynamics is sometimes discussed in awestruck, almost mystical terms.

Keeping in mind some of the ideas which have been discussed about the relationships between naming, categorising and counting, my own instinct is that the missing space in the diagram we just saw should relate to some (as yet unacknowledged?) aspect of *consciousness*. We encountered hints in Chapter 31 that the physical creation of a Riemann dynamics might lead to entirely new physics and, consequently, new technology. This is all highly speculative, of course, but the possibility (however faint) that a new technology might emerge which somehow involved or interfaced with consciousness should set alarm bells ringing! It seems evident that our collective wisdom has not kept pace with our technological evolution in recent centuries, and this would be a situation requiring *considerable* wisdom. In the face of this possibility, then, and considering its record of technological abuse and misuse, Western civilisation would do well to *tread lightly*.

notes

A RE-REINTRODUCTION [pages 1–2]

1. M. Lapidus, *In Search of the Riemann Zeros* (AMS, 2008) p. xiv

CHAPTER 28 [pages 3–24]

1. named after French mathematician Charles Hermite (1822–1901)

2. In Thomas Pynchon's wonderfully sprawling historical novel *Against the Day* (2006), the fictional character Yashmeen Halfcourt is revealed as the muse responsible for giving Hilbert this idea, with a question she asks from the back of one of his lectures while visiting Göttingen from Cambridge.

3. from a letter to Andrew Odlyzko, dated January 3, 1982 [http://www.dtc.umn.edu/~odlyzko/polya/index.html also archived as http://tinyurl.com/dyn3huq]

4. *ibid*

5. J. Derbyshire, *Prime Obsession* (John Henry Press, 2003) p. 351

6. see http://empslocal.ex.ac.uk/people/staff/mrwatkin/zeta/physics4.htm [also available as http://tinyurl.com/cppaa4f]

7. At the time of writing this, I understand only a small part of the current theory surrounding the Selberg Trace Formula, but enough to glimpse the beauty.

8. These were inspired by Escher's series of wood engravings *Circle Limit I–IV*.

9. Although I've borrowed this explanation from somewhere, I'm unable to recall the original source. However, several explanations involving "shrinking rulers" can be found in the literature, and they've become fairly standard in popular accounts of Einstein's theory of special relativity (which also involves hyperbolic "metrics").

10. Escher had no formal mathematical training, his approach to geometry being almost entirely intuitive and visual. However, this particular series was inspired by his discussions with the Canadian mathematician H.S.M. Coxeter in the 1950s.

11. As the *Euler characteristic* of a torus equals zero, the *Gauss–Bonnet theorem* tells us that it can't have a metric of constant negative curvature, so it can't possibly be expressed a "quotient"

of the hyperbolic disc (which has such a metric). See `http://en.wikipedia.org/wiki/Gauss-bonnet` [also available as `http://tinyurl.com/a4g82e9`].

12. A. Weil, "Sur les 'formules explicites' de la théorie des nombres premiers", *Meddelanden Från Lunds Univ. Mat. Sem.* (1952) pp. 252–265. More details on the explicit formula can be found at `http://empslocal.ex.ac.uk/people/staff/mrwatkin/zeta/weilexplicitformula.htm` [also available as `http://tinyurl.com/bvtlnc6`].

13. Although Hilbert and Pólya (separately) introduced the idea of the spectral interpretation, this was more like wishful thinking than "evidence".

14. Subtle differences between the forms of the Selberg Trace Formula and the Riemann–Weil explicit formula make this a mathematical impossibility. In D. Goldfeld's "Explicit formulae as trace formulae" [in *Number Theory, Trace Formulas and Discrete Groups* (K.E. Aubert, *et al.*, eds.) (Academic, 1989) pp. 281–288], he shows "*that Weil's explicit formula can in fact be interpreted as a trace formula on a suitable space*", but this suitable space is far more involved than a compact Riemann surface.

15. See P.D. Lax and R.S. Phillips, *Scattering Theory for Automorphic Functions* (Princeton University Press, 1976) which developed ideas in L.D Faddeev and B.S. Pavlov, "Scattering theory and automorphic functions", *Seminar of Steklov Mathematical Institute of Leningrad* 27 (1972) pp. 161–193.

16. Non-chaotic systems are also called "integrable", although this shouldn't be confused with the more familiar sense in which that word is used in elementary calculus (there is an indirect relationship, but it's quite a complicated one).

CHAPTER 29 [pages 25–34]

1. The probability distribution on the space of $n \times n$ Hermitian matrices is described by the function

$$\frac{1}{2^{n/2}\pi^{n^2/2}}e^{-\frac{n}{2}\mathrm{tr}H^2}$$

where H is a typical matrix and $\mathrm{tr}\,H$ its *trace* (the sum of its diagonal elements).

2. E.P. Wigner, "Characteristic vectors of bordered matrices with infinite dimensions", *Annals of Mathematics* 62 (1955) pp. 548–564

3. This was primarily in optics and electronics, although their role was fairly marginal.

4. This is Montgomery's recollection as recounted to John Derbyshire in *Prime Obsession* (John Henry Press, 2003) pp. 287–288.

5. Although still unproved, Montgomery's pair correlation conjecture has been extended (as a plausible conjecture) beyond pairs to general "*n*-tuples" of zeros, as well as to zeta functions associated with *automorphic representations*. Also, an analogous conjecture for *L*-functions (see Chapter 25) *was* proved in 1982 by Montgomery's student A.E. Özlük.

6. Although Odlyzko has calculated hundreds of millions of Riemann zeros, he hasn't focussed on the initial stretch from 0 up the critical line, but rather (and more interestingly) has looked at the intricacies of what's going on with the zeros much "higher up". This explains why his name doesn't appear in the table that appeared in Chapter 21.

7. J.P. Keating and N.C. Snaith, "Random matrix theory and $\zeta(1/2 + it)$", *Communications in Mathematical Physics* 214 (2000) pp. 57–89

CHAPTER 30 [pages 35–51]

1. J.A. Yorke and T.Y. Li , "Period three implies chaos", *American Mathematical Monthly* 82 (1975) pp. 985–992. This paper also introduced the use of the adjective "chaotic" in the same context.

2. M.C. Gutzwiller, "Periodic orbits and classical quantization conditions", *Journal of Mathematical Physics* 12 (1971) pp. 343–358. Incidentally, the term *trace* originates with matrix mathematics, where the trace of a matrix is simply the sum of its diagonal entries. This concept generalises from matrices to operators, so it's still possible to define the trace of operators on infinite-dimensional spaces (where no matrix representation would be possible). Both Selberg's and Gutzwiller's trace formulas involve traces of operators.

3 It's possible to describe a very simple motion on the hyperbolic disc which is chaotic and which can be described in terms of Gutzwiller's formula. Compact Riemann surfaces can be constructed by "curling up" the hyperbolic disc as I attempted to explain in Chapter 28, so a corresponding motion on such surfaces can be similarly described. Applied to such a dynamical system, the GTF "collapses down to" (takes the same form as) the STF. See M.C. Gutzwiller, "Classical quantization of a Hamiltonian with ergodic behavior", *Physical Review Letters* 45 (1980) pp. 150–153. It should be noted that, in general, Gutzwiller's formula is only intended to work at the level of approximation (it's *asymptotic*) but in this special case it becomes (like Selberg's formula) an *exact* relation. Incidentally, the fact that general trace formulas are extensions of Selberg's had earlier been noted by Colin de Verdière [*Comptes Rendus Acad. Sci. Paris* 276 (1973) p. 1518].

4. John Derbyshire, *Prime Obsession* (John Henry Press, 2003) p. 291

5. This is something of an oversimplification. There is no such thing as a "GUE matrix" – rather, the GUE provides a probability distribution over the space of all Hermitian matrices of a given size. Imagine a "random matrix generator" using this distribution – any Hermitian matrix of the given size can show up, but some matrices have a much greater chance of showing up than others.

Consequently, certain tendencies in the spectra of the "output" matrices will start to manifest, statistically. Changing the random matrix ensemble (to the GOE, say) will lead to different spectral tendencies. It's the GUE tendencies which the zeta zeros appear to be displaying.

6. More accurately, the meaning of "classical physics" depends on the context. When discussing relativistic physics (rather then quantum mechanics), it refers to Newtonian physics, the physics which preceded relativity theory.

7. quoted in J. Brown, "Where two worlds meet", *New Scientist* (18 May 1996) p. 28

8. *Charge* is most directly familiar as the static electricity which occasionally gives you a minor shock. *Action* is a key quantity in physics, defined as energy × time (so typically measured in "Joule-seconds"). George Stoney discovered the quantisation of charge in 1881, which led to the now largely obsolete *Stoney units* based on his discoveries. In 1889, Max Planck discovered the quantisation of *action*, and the better known *Planck units* were adopted according to his scheme.

9. Water forms into drops because of a physical phenomenon called *surface tension*. I must stress that I'm not trying to relate this to quantum mechanics, just referring to a familiar phenomenon as part of an analogy.

10. Bolzmann's constant is associated with thermodynamics. The Coulomb constant k_e is associated with electrostatics, equalling $1/4\pi\varepsilon_0$, where ε_0 is the *permittivity of free space*. The gravitational constant G is a universal one, not tied to the gravity here on Earth. Planck's constant, incidentally, is 6.626×10^{-34} Joule-seconds.

11. One Planck mass, recall, is about 22 millionths of a gram. One Planck energy is approximately 500 kilowatt-hours, the sort of quantity associated with heavy industry. This allows us to see how "$E = mc^2$" implies that an awful lot of energy is locked up in a small amount of matter.

12. R. Penrose, *The Emperor's New Mind* (Oxford University Press, 1989), p. 348

13. http://en.wikipedia.org/wiki/Schrodinger_equation

14. http://en.wikipedia.org/wiki/Uncertainty_principle

15. L.D. Fadeev, "What is complete integrability in quantum mechanics?", *AMS Translations (2)* 220 (2007) p. 83

16. P. Bourgade and J.P. Keating, "Quantum chaos, random matrix theory, and the Riemann ζ-function", *Séminaire Poincaré* XIV (École Polytechnique, 2010) p. 130

17. D. Goldfeld, "Explicit formulae as trace formulae", in *Number Theory, Trace Formulas and Discrete Groups*, K.E. Aubert, *et al.*, eds. (Academic, 1989) pp. 281–288

18. P. Bourgade and J.P. Keating, "Quantum chaos, random matrix theory, and the Riemann ζ-function", *Séminaire Poincaré* XIV (École Polytechnique, 2010) p. 130

19. O. Bohigas, M.-J. Giannoni and C. Schmit, "Characterisation of chaotic quantum spectra and universality of level fluctuation laws", *Physical Review Letters* 52 (1984) pp. 1–4

20. More on time-reversible and -irreversible systems can be found in J.S.W. Lamb and J.A.G. Roberts, "Time-reversal symmetry in dynamical systems: A survey", *Physica D* 112 (1998) pp. 1–39.

21. D. Zagier, "The first 50 million prime numbers", *The Mathematical Intelligencer* 0 (1977) pp. 7–19

22. In *Quantum Chaos and Statistical Nuclear Physics* (eds. T.H. Seligman and H. Nishioka), Lecture Notes in Physics 263 (Springer, 1986) pp. 1–17

23. L.D Faddeev and B.S. Pavlov, "Scattering theory and automorphic functions", *Seminar of Steklov Mathematical Institute of Leningrad* 27 (1972) pp. 161–193

24. M.C. Gutzwiller, "Stochastic behavior in quantum scattering", *Physica D* 7 (1983) pp. 341–355

25. On p. 8 of Berry's paper he explains that "*An apparently anomalous outcome of Odlyzko's computation of the form factor K(τ) of the Riemann zeros gives further support to the semiclassical analogy. Although he finds good overall agreement with the GUE formula, close examination of the difference K–K$_{GUE}$ reveals a series of spikes for small τ. Such spikes are predicted by the semiclassical theory, because…the universality of the GUE statistics ceases to hold for large energy scales, that is short time scales.*" Incidentally, the *form factor* is the Fourier transform of the "pair correlation function" mentioned on p. 32. Berry shows how these spikes correspond to logarithms of powers of primes, thereby strengthening the semiclassical interpretation of the zeta zeros.

26. M.V. Berry and J.P. Keating, "*H = xp* and the Riemann zeros", in *Supersymmetry and Trace Formulae: Chaos and Disorder*, ed. I.V. Lerner, *et al.* (Kluwer, 1999) pp. 355–367

27. see http://empslocal.ex.ac.uk/people/staff/mrwatkin/zeta/physics1.htm [also available as http://tinyurl.com/clccjk5]

28. M.V. Berry and J.P. Keating, "The Riemann zeros and eigenvalue asymptotics", *SIAM Review* 41 (1999) pp. 236–266

29. M. du Sautoy, *The Music of the Primes: Why an Unsolved Problem in Mathematics Matters* (HarperCollins, 2003) p. 280

CHAPTER 31 [pages 53–63]

1. It could be argued that this isn't a unique phenomenon, by pointing to, for example, Michael

Atiyah's work wherein deep results in algebraic topology and *K*-theory (that's pure mathematics) have been arrived at via consideration of ideas from areas of physics such as quantum field theory and string theory. My description of the situation is perhaps oversimplified, as there is much traffic back and forth between abstract mathematics and physics. However, the fact that this instance concerns the *very foundations of the number system* (rather than some exotic feature in the mathematical landscape which is built thereupon) suggests to me that there is something *profoundly* singular about it.

2. G. Tenenbaum and M. Mendès France, *The Prime Numbers and Their Distribution* (AMS, 2000) p. 1

3. N. Snaith, "Random matrix theory and zeta functions" (Ph.D. thesis, Bristol University, 2000)

4. E. Bombieri, "Prime territory: Exploring the infinite landscape at the base of the number system", *The Sciences* 32 (1992) p. 36

5. J. Brown, "Where two worlds meet", *New Scientist* (18 May 1996) p. 30

6. J. Derbyshire, *Prime Obsession* (Joseph Henry Press, 2003), p. 360

7. A. Connes, "Formule de trace en géométrie non commutative et hypothèse de Riemann", *C.R. Sci. Paris* 323 (1996) pp. 1231–1235

8. quoted in E. Klarreich, "Prime time", *New Scientist*, 11 November 2000, p. 36

9. P. Davies, "Reality in the melting pot", *The Guardian*, 23 September 2003 [http://www.guardian.co.uk/science/2003/sep/23/spaceexploration.comment, also available as http://tinyurl.com/5786yk]

10. S.R. Beane, Z. Davoudi and M.J. Savage, "Constraints on the universe as a numerical simulation" (preprint, 2012) [http://arxiv.org/abs/1210.1847]

CHAPTER 32 [pages 65–81]

1. The first problem (algebra) is the *Brumer–Stark Conjecture* [http://www.wikipedia.org/wiki/Brumer-Stark_conjecture]. The second (geometry) is *Hadwiger's Conjecture* [http://www.wikipedia.org/wiki/Hadwiger_conjecture_(combinatorial_geometry)]. The third (analysis) is *Khabibullin's Conjecture* [http://www.wikipedia.org/wiki/Khabibullin's_conjecture_on_integral_inequalities].

2. P. Hoffman, *Archimedes Revenge: The Joys and Perils of Mathematics* (Fawcett Crest, 1988) pp. 36–37

3. A much-cited example of this is Andrew Wiles' research which led to a proof of Fermat's Last Theorem in the 1990s. Although this theorem concerns the interrelations of positive integers,

Wiles' research led to major progress in the theory of *elliptic curves*.

4. G.H. Hardy, *Collected Papers,* vol. II (Oxford University Press, 1967) p. 14

5. quoted in W. Sartorius von Waltershause, *Gauss: zum Gedächtniss* (Hirzel, 1856) p. 79

6. J.P. Keating, "Physics and the Queen of Mathematics", *Physics World* (April 1990) p. 50

7. Additive number theory typically deals with such questions as how many different ways a given counting number can be expressed as a sum of counting numbers, or squared counting numbers, or primes (problems of "additive partitions"): see http://empslocal.ex.ac.uk/people/staff/mrwatkin/zeta/partitioning.htm.

8. S. Hawking, "Zeta function regularization of path integrals in curved spacetime", *Communications in Mathematical Physics* 55 (1977) pp. 133–148. A precedent for this use of zeta function regularisation appeared a year earlier in J.S. Dowker and R. Critchley, "Effective Lagrangian and energy-momentum tensor in de Sitter space", *Physical Review D* 13 (1976) pp. 3224–3232. Hendrik Casimir's work on vacuum energy in the late 1940s arguably involves the earliest application of the technique.

9. see http://en.wikipedia.org/wiki/Lee-Yang_theorem [also available as http://tinyurl.com/crj23lp]

10. B.L. Julia, "Statistical theory of numbers", in *Number Theory and Physics*, J.M. Luck, *et al.,* eds. (Springer-Verlag, 1990) pp. 276–293

11. Also known as a *Hagedorn singularity*. Julia explains *"one can reinterpret the pole of the zeta function at s = 1 as a breakdown of the grand-canonical ensemble at the Hagedorn temperature because of the exponential increase of the density of energy levels for large energies"*, referencing R. Hagedorn, "Statistical thermodynamics of strong interactions at high energies", *Suppl. Nuovo Cimento* 3 (1965) pp. 147–186, as well as Yu. B. Rumer, *Zh. Eksp. Teor. Fiz.* 38 (1960) p. 1899

12. see http://en.wikipedia.org/wiki/Kramers-Wannier_duality [also available as http://tinyurl.com/d4veeo9]

13. G.W. Mackey, *Unitary Group Representations in Physics, Probability and Number Theory* (Benjamin, 1978), pp. 299, 321

14. D. Spector, "Supersymmetry and the Möbius inversion function", *Communications in Mathematical Physics* 127 (1990) pp. 239–252

15. References to Andreas Knauf's work, as well as that of several of his collaborators, can be found at http://empslocal.ex.ac.uk/people/staff/mrwatkin/zeta/spinchains.htm [also available as http://tinyurl.com/dx35tqy].

16. D. Fivel, "The prime factorization property of entangled quantum states" (preprint, 1994) [available at http://arxiv.org/abs/hep-th/9409150]

17. see http://en.wikipedia.org/wiki/Pauli_exclusion_principle

18. M.J. Shai Haran, *The Mysteries of the Real Prime* (Oxford University Press, 2001)

19. M.V. Berry and J.P. Keating, "The Riemann zeros and eigenvalue asymptotics", *SIAM Review* 41 (1999) pp. 236–266

20. W. Parry and M. Pollicott, "An analogue of the prime number theorem for closed orbits of axiom A flows", *Annals of Mathematics* 118 (1983) pp. 573–591

21. J.H. Hannay and A.M. Ozorio De Almeida, "Periodic orbits and a correlation function for the semiclassical density of states", *Journal of Physics A* 17 (1984) pp. 3429–3440. The connection with the PNT was later observed by J. Keating in "Physics and the Queen of Mathematics", *Physics World* (April 1990).

22. see http://empslocal.ex.ac.uk/people/staff/mrwatkin/zeta/physics.htm [also available as http://tinyurl.com/83yldt9]

23. see http://empslocal.ex.ac.uk/people/staff/mrwatkin/zeta/physics.htm [also available as http://tinyurl.com/83yldt9]

24. H. Mack, M. Bienert, F. Haug, M. Freyberger and W.P. Schleich, "Wave packets can factorize numbers", *Physica Status Solidi* (B) 233 (2002) pp. 408–415

25. see http://empslocal.ex.ac.uk/people/staff/mrwatkin/zeta/dynamicalNT.htm [also available as http://tinyurl.com/osng5ze]

26. quoted in K. Sabbagh, *Dr. Riemann's Zeros* (Farrar, Strauss & Giroux, 2003) pp. 267–268

27. B. Cipra, "A prime case of chaos", *What's Happening in the Mathematical Sciences* 4 (American Mathematical Society, 1999) p. 35

CHAPTER 33 [pages 83–111]

1. Recall that this spiral's crossings, travelling out from an initial crossing at 1, are found at 2.718..., 2.718... × 2.718... and higher powers. In mathematical notation, the spiral is given by $r = e^{\theta/2\pi}$.

2. T. Gowers, *Mathematics: A Very Short Introduction* (Oxford University Press, 2002) p. 121

3. It could be argued that the "foundational level of mathematics" would be that area of study

commonly known as the *foundations of mathematics* (dealing with axiomatic systems, logic and sets), but I would point out that this is more correctly understood as part of *philosophy*.

4. M. du Sautoy, *The Music of the Primes: Why an Unsolved Problem in Mathematics Matters* (HarperCollins, 2003) pp. 6–7

5. This appeared in D. Mackenzie, "Homage to an itinerant master", *Science* 275 (1997) p. 759, wrongly attributed to the prolific number theorist Paul Erdős. While addressing a memorial seminar in San Diego shortly after Erdős died, Myerson proposed it as something he (and Kac – see Chapter 22) *might have* said in response to Einstein's famous remark, but it somehow went into circulation as something he *did* say. See `http://empslocal.ex.ac.uk/people/staff/mrwatkin/kac-pomerance.txt` [also available as `http://tinyurl.com/c3v4fxc`].

6. D. Zagier, "The first 50 million prime numbers", *The Mathematical Intelligencer* 0 (1977) p. 7

7. G. Spencer-Brown, *Probability and Scientific Inference* (Longmans, Green & Co., 1957) p. 105

8. R.C. Vaughan, quoted in A. Granville, "Harald Cramér and the distribution of prime numbers", *Scandinavian Actuarial Journal* 1995 (1995) p. 12

9. See, for example, Chapter 2 of Donald Knuth's *Things a Computer Scientist Rarely Talks About* (CSLI, 2003). Knuth is an esteemed computer scientist at Stanford University as well as a practicing Christian.

10. This method was developed by the Switzerland-based engineer John Walker. Some theorists have argued passionately that only three emissions are required to generate each 0 or 1 (comparing time intervals between first and second, second and third), others have argued just as passionately that this would lead to subtle biases. No one, though, has questioned the "true randomness" of the output with Walker's four-emissions version. One minor detail that needs to be considered here is *source decay* – the radioactive material very gradually becomes less radioactive, leading to the gaps between successive emissions very gradually getting slightly larger on average. Walker has demonstrated conclusively that for the time scales his method would normally operate on to generate any reasonable quantity of random data, this effect would be irrelevant. See `http://www.fourmilab.ch/hotbits`.

11. U. Dudley, *Elementary Number Theory*, 2nd edition (W.H Freeman & Company, 1978) p. 163

12. T. Gowers, *Mathematics: A Very Short Introduction* (Oxford University Press, 2002) p. 118

13. J. Muir, *Of Men and Numbers: The Story of the Great Mathematicians* (Dover Publications, 1996) p. 6

14. E. Bombieri, "Prime territory: Exploring the infinite landscape at the base of the number system", *The Sciences* 32 (1992) p. 36

15. M. du Sautoy, *The Music of the Primes: Why an Unsolved Problem in Mathematics Matters* (HarperCollins, 2003) p. 45

16. A. Doxiadis, *Uncle Petros and Goldbach's Conjecture* (Bloomsbury, 2000) p. 84

17. E. Klarreich, "Prime time", *New Scientist* (11 November 2000) p. 36

18. If there's an imbalance in the odd/even factor count then the graph will stray outside the "power curves", the "coin" we're looking at wouldn't be a fair one and the RH would be false. There is a subtlety which was passed over, though. The Mertens function doesn't just take the values +1 and −1, it can also take the value 0. However, if we let s_n denote the nth "square-free" integer and define $M^*(n) = M(s_n)$, where $M(n)$ is the Mertens function (see pp. 157–161 of Vol. 2), then we know $s_n \sim \pi^2 n/6$ (see p. 48 of Vol. 2), so if the growth of M is $O(n^{1/2+\varepsilon})$ (see pp. 151–156 of Volume 2) then so is that of M^*, and *vice versa*.

19. G. Tenenbaum and M. Mendès France, *The Prime Numbers and Their Distribution* (American Mathematical Society, 2000) p. xii

20. *ibid*, p. 51

21. A. Connes and M. Marcolli, "From physics to number theory via noncommutative geometry. Part I: Quantum statistical mechanics of Q-lattices" (2004 preprint) [http://arxiv.org/abs/math/0404128, also available as http://tinyurl.com/d8qprtt]

22. In order to understand what adeles are, you first need to understand p-adic number systems, which will be examined in Chapter 34. There's a p-adic number system for each prime number p, and an adele (roughly speaking) is a numerical entity with infinitely many components: one p-adic number for each prime, together with a single real number.

23. E. Klarreich, "Prime time", *New Scientist* (11 November 2000) p. 36

24. M. Schroeder, *Number Theory in Science and Communication* (Springer, 1984) p. 307

25. P. Ribenboim, *The Book of Prime Number Records* (Springer, 1988) p. 153

26. E. Bombieri, "Prime territory: Exploring the infinite landscape at the base of the number system", *The Sciences* 32 (1992) p. 36

27. J. Brian Conrey, "The Riemann Hypothesis", *Notices of the American Mathematical Society* (March 2003) p. 345

28. quoted in K. Sabbagh, *Dr. Riemann's Zeros* (Farrar, Strauss & Giroux, 2003) p. 268

29. quoted in *ibid*, p. 269

30. M. du Sautoy, *The Music of the Primes: Why an Unsolved Problem in Mathematics Matters* (HarperCollins, 2003) p. 55

31. This would be mathematically represented as the mapping $T \to (T/2\pi)\ln(T/2\pi)$.

32. see `http://en.wikipedia.org/wiki/Poisson_distribution` [also available as `http://tinyurl.com/9zg9d`]

33. see `http://empslocal.ex.ac.uk/people/staff/mrwatkin/zeta/physics6.htm` [also available as `http://tinyurl.com/ctam2ms`]

34. T. Gowers, *Mathematics: A Very Short Introduction* (Oxford University Press, 2002) p. 121

CHAPTER 34 [pages 113–127]

1. *The Confessions of Saint Augustine*, Book XI, translated by E.B. Pusey (Cosimo, 2006)

2. "The Nature of Time: Geometry, Physics & Perception" was held at Tatranská Lomnica in the Slovak Republic, May 2002), and since 1984 there has been an annual Russian Interdisciplinary Temporology Seminar [`http://www.chronos.msu.ru/seminar/eindex.html`, also archived as `http://tinyurl.com/bpsdh4b`].

3. The English version of their website site `http://www.chronos.msu.ru` [also archived as `http://tinyurl.com/bv3c73q`] translates the Russian name as the rather more awkward "Institute for Time Nature Explorations".

4. Conferences under the name "Toward a Science of Consciousness" have been held bianually at the University of Arizona, Tucson since 1994, and that university now hosts a Center for Consciousness Studies [`http://www.consciousness.arizona.edu`].

5. The *Journal of Consciousness Studies* (ISSN 1355-8250) has been published monthly since 1994 and is affiliated with the conferences and Center mentioned in Note 4 above.

6. The Tibetan and Zen Buddhist traditions have extensively investigated the nature of consciousness, but this is a mediation-based approach, differing significantly from the rational/conceptual thought-based approach of Western science.

7. J.J. Sylvester, "On certain inequalities relating to prime numbers", *Nature* 38 (1888) pp. 259–262, reproduced in *Collected Mathematical Papers*, Vol. 4 (Chelsea, 1973) p. 600

8. *Quaternions*, for example, are a type of four-dimensional number, first proposed by William Hamilton in 1843 [`http://en.wikipedia.org/wiki/Quaternion`]. Similarly, it is possible to study eight-dimensional *octonions* [`http://en.wikipedia.org/wiki/Octonions`].

9. *cf.* R. Calvert and D. Brock, "Silver Machine" (EMI Music Publishing, 1972)

10. The use of complex numbers to represent a kind of two-dimensional "complex time" can occasionally be found in the quantum mechanics literature. See also K. Scharff's bibliography of alternative time models [http://www.fourmilab.ch/rpkp/scharff.html, also archived as http://tinyurl.com/c2qws9w].

11. N. Tomalin and R. Hall, *The Strange Voyage of Donald Crowhurst* (Penguin, 1973), pp. 256 and 262–263. Crowhurst had previously been non-religious. His use of the word "imaginary" here is certainly informed by a familiarity with complex numbers. He had a background in electronics and his log contained philosophical musings on the mystery of the imaginary unit *i*.

12. see http://en.wikipedia.org/wiki/Ramanujan_summation

13. see D.H. Lehmer, "Incomplete Gauss sums", *Mathematika* 23 (1976) pp. 125–135

14. See http://en.wikipedia.org/wiki/Mobius_function. We met the Möbius function in Chapter 24 (although it wasn't named as such).

15. See http://en.wikipedia.org/wiki/Von_Mangoldt_function. We met the von Mangoldt function in Chapter 12 (although it wasn't named as such).

16. M. Planat, "On the cyclotomic quantum algebra of time perception", *Neuroquantology* 2 (2004) pp. 292–308

17. I.V. Volovich, "Number theory as the ultimate physical theory", *p-Adic Numbers, Ultrametric Analysis and Applications* 2 (2010) pp. 77–87

18. The First International Conference on *p*-Adic Mathematical Physics was held at the Steklov Institute in Moscow in 2003. Subsequent conferences have been held in 2005, 2007 and 2009.

19. On p. 80 of his paper "Number theory as the ultimate physical theory" [*p-Adic Numbers, Ultrametric Analysis and Applications* 2 (2010)], Volovich observes that unlike the field of real numbers ℝ, the field of *p*-adic numbers lacks an intrinsic "greater than/less than" sense of order, relating this to discussions by Penrose and Hawking on the "arrow of time".

20. The areas of physics in which *p*-adic and adelic mathematics have proved most useful are string theory and quantum field theory, although quite a lot of work has been done in other areas. See http://empslocal.ex.ac.uk/people/staff/mrwatkin/zeta/physics7.htm [also available as http://tinyurl.com/c637edc].

CHAPTER 35 [pages 129–139]

1. quoted in K. Sabbagh, *Dr. Riemann's Zeros* (Farrar, Straus and Giroux, 2003) p. 189

2. quoted in *ibid*, p. 246

3. S. Dehaene, *The Number Sense: How the Mind Creates Mathematics* (Oxford University Press, 1997)

4. L. Kauffman, "Virtual logic – formal arithmetic", *Cybernetics & Human Knowing* 7 (2000) pp. 91–95 [see http://www.imprint.co.uk/C&HK/vol7/kauffman_7-4.pdf, also available as http://tinyurl.com/nnfqtp].

5. A. Doxiadis, *Uncle Petros and Goldbach's Conjecture* (Bloomsbury, 2000) pp. 183–184

6. We could even say "counting^2" or "counting squared".

7. I was surprised to find that this kind of device goes back to the very earliest animated films, occurring regularly in Max Fleischer's *Out of the Inkwell* (1918–1929), featuring Koko the Clown.

8. quoted in K. Sabbagh, *Dr. Riemann's Zeros* (Farrar, Straus and Giroux, 2003) p. 246

9. This used to be part of a standard undergraduate curriculum, although with the decline of standards in Western university-level mathematics, it isn't commonly taught these days. Basically, the real numbers are defined as a set of *equivalence classes* of *Cauchy sequences* of rational numbers. For details, see http://en.wikipedia.org/wiki/Tarski's_axiomatization_of_the_reals [also available as http://tinyurl.com/c7k42b5].

CHAPTER 36 [pages 141–155]

1. J.-B. Bost and A. Connes, "Hecke algebras, type III factors and phase transitions with spontaneous symmetry breaking in number theory", *Selecta Mathematica (N.S.)* 1 (1995) pp. 411–457

2. A. Connes, "Noncommutative geometry year 2000", *Highlights of Mathematical Physics*, ed. A.S. Fokas, *et al.* (AMS, 2002) pp. 49–110 [available as http://www.arxiv.org/abs/math/0011193]

3. J.-B. Bost and A. Connes, "Hecke algebras, type III factors and phase transitions with spontaneous symmetry breaking in number theory", *Selecta Mathematica (N.S.)* 1 (1995) pp. 411–457

4. However well engineered and to whatever level of precision, the accuracy of a physical realisation of a mathematical model of a classical, macroscopic system will always break down as the scale of our scrutiny extends down to the atomic level of matter. This is because of our false assumptions about a continuity that (as with the real number system) continues to operate at any scale, however small.

5. J.J. Sylvester, "On certain inequalities relating to prime numbers", *Nature* 38 (1888) pp. 259–262, reproduced in *Collected Mathematical Papers*, Vol. 4 (Chelsea, 1973) p. 600

6. I define *scientism* as the belief that science is the exclusive and complete route to truth. To be *scientistic* is to adhere to scientism. See R. Sheldrake, *The Science Delusion* (Coronet, 2012)

7. C. Deninger, "Some ideas on dynamical systems and the Riemann zeta function" (preprint from proceedings of the 1997 ESI conference on the Riemann zeta function); "Some analogies between number theory and dynamical systems on foliated spaces", *Documenta Mathematica*, Extra Volume ICM I (1998) pp. 163–186; "Number theory and dynamical systems on foliated spaces", *Jahresbericht der Deutschen Mathematiker-Vereinigung* 103 (2001) pp. 79–100

8. M. Lapidus, *In Search of the Riemann Zeros* (AMS, 2008)

9. A. Watts, *Tao: The Watercourse Way* (Pantheon, 1975) p. 15

10. D. Wade, *Li: Dynamic Form in Nature* (Wooden Books, 2003)

11. A. Watts, *The Wisdom of Insecurity* (Pantheon, 1951) p. 41

12. see `http://empslocal.ex.ac.uk/people/staff/mrwatkin/zeta/NTfractality.htm` [also available as `http://tinyurl.com/bosh7h5`]

CHAPTER 37 [pages 157–172]

1. G. Tenenbaum and M. Mendès France, *The Prime Numbers and Their Distribution* (AMS, 2000) p. 1

2. see Appendix 16

3. Transpersonal psychology can be thought of as *"the area of psychology that focuses on the study of transpersonal experiences and related phenomena. These phenomena include the causes, effects and correlates of transpersonal experiences and development, as well as the disciplines and practices inspired by them"*, where "transpersonal experiences" are defined as *"experiences in which the sense of identity or self extends beyond (trans) the individual or personal to encompass wider aspects of humankind, life, psyche or cosmos."* [R. Walsh and F. Vaughan, "On transpersonal definitions", *Journal of Transpersonal Psychology* 25 (1993) p. 203]

4. There is a subtlety here: as was touched on in Chapter 3 in the section on "partitions" of counting numbers (also Chapter 32, note 7), *additive number theory* can lead to some incredibly difficult problems and results. But as explained there, these issues, like multiplication, involve the "turning of the number system on itself" (counting not objects, but *numbers themselves*, or in the case of partitions, *ways of combining numbers*).

5. Marie-Louise von Franz, *Number and Time: Reflections Leading Toward a Unification of Depth Psychology and Physics* (Northwestern University Press, 1974). This book is problematic in some ways – von Franz's lack of familiarity with higher mathematics becomes apparent in several places, leading to dubious assertions and conclusions. Nevertheless, it contains some fascinating insights. To be read with appropriate caution.

6. J. Andersson, "On generalized Hardy classes of Dirichlet series", 2012 preprint [see `http://arxiv.org/abs/1207.5337`]

7. A. Huxley, "A visionary prediction" (1963), reprinted in P. Haining, ed., *The Hashish Club: An Anthology of Drug Literature* Vol. 1 (Peter Owen, 1975) p. 152

8. *ibid*, p. 154

9. This translation is due to Judith Simmer-Brown, from "Madhyamaka and the prasaṅga method: A pedagogy for working with conceptual mind" (unpublished conference proceedings, 1981).

CHAPTER 38 [pages 173–181]

1. "Sacred geometry" usually involves such things as the Golden Ratio (see Chapter 9), the five *Platonic solids* and simple compass-and-straightedge constructions such as the *vesica piscis* and the so-called "Flower of Life". I've never really liked the term as I can't see why some aspects of geometry are more "sacred" than others – either it's all sacred or none of it is.

2. L.L. Whyte's concept of "European dissociation" may be relevant to this discussion – see his *The Next Development in Man* (Holt, 1948) [later republished as *The Next Development in Mankind*]. The much more widely known author Ken Wilber dwells on Whyte's work at great length in his *Up From Eden: A Transpersonal View of Human Evolution* (Doubleday, 1981).

3. Slavery, for example, certainly hasn't been unique to the West. The enslavement of people, stealing of resources and making extinct of species were not categorically new behaviour – it's the unprecedented extent which I'm stressing here.

4. I have no anthropological qualifications and I'm fully aware of the dangers of romanticising so-called primitive cultures. But, *still*, it just doesn't seem to me that the indigenous cultures of, say, Australia, Southern Africa or the Amazon basin (those being among the least "numerate") were on an evolutionary trajectory that would have led them to resemble anything like that of the modern West.

5. These decisions are almost always being made by men in Western-style suits and ties – strange, isn't it?

6. There are probably still a few people around in 2013 who would disagree with that, but (other than the truths of number theory) there's very little you can say that *someone* isn't going to take issue with.

7. Marie-Louise von Franz, *Number and Time: Reflections Leading Toward a Unification of Depth Psychology and Physics* (Northwestern University Press, 1974). This book is in some ways problematic – the author's lack of familiarity with higher mathematics becomes apparent in several places, leading to questionable conclusions and assertions. Still, it contains some fascinating insights. Read with appropriate caution.

8. Guénon became something of a religious fundamentalist, and I wouldn't endorse his overall worldview, but his book *The Reign of Quantity and the Signs of the Times* (Luzac, 1953) contains much interesting insight into Westernised humanity's relationship with number.

9. Curiously, this word which I came up with while putting together Volume 1 was separately coined around the same time in J.L.M. McCoyd *et al.*, "Quantocentric culture: Ramifications for social work education", *Social Work Education* 28 (2009) pp. 811–827. The authors of that paper define "quantocentric culture" as that in which *"quantitative research methods are privileged over other lines of inquiry"*.

10. Note that there's not been a *complete* rejection, considering the ongoing popularity of numerology, the absence of thirteenth floors in many Western hotels, *etc.* And outside the Western mainstream there has always been a marginal current associated with the so-called "perennial philosophy" which is open to the qualitative approach to number – see, for example, Aldous Huxley's *The Perennial Philosophy* (Harper, 1945).

11. David Abram's *The Spell of the Sensuous* (Vintage, 2007) deals with these issues and I'd highly recommend it – his humility in not claiming to have all the answers is refreshing. Leonard Shlain's *The Alphabet Versus the Goddess* (Penguin, 1998) is also worth a read.

12. "spirit of the age" (literally "time spirit" in German), which is discussed as if it had a life of its own, revealing itself in fashion and design trends, changing social attitudes and cultural obsessions

CHAPTER 39 [pages 183–199]

1. A. Knauf, "Number theory, dynamical systems and statistical mechanics" *Reviews in Mathematical Physics* 11 (1999) p. 1027

2. T. Tao, "Every odd number greater than 1 is the sum of at most five primes" (preprint, 2012) [see http://arxiv.org/abs/1201.6656]

3. M. du Sautoy, "The music of the primes", *Science Spectra* 11 (1998)

4. D. Rockmore, "Chance in the primes", *Chance News* 10 (2002) [http://www.dartmouth.edu/~chance/chance_news/recent_news/chance_news_10.10.html, also available as http://tinyurl.com/c4b92nw]

5. G. Tenenbaum and M. Mendès France, *The Prime Numbers and Their Distribution* (American Mathematical Society, 2000) p. 44

6. J. Brown, "Where two worlds meet", *New Scientist* (18 May 1996) p. 30

7. J.P. Keating, "Physics and the Queen of Mathematics", *Physics World* (April 1990) p. 50

8. M.V. Berry, "Chaos and the semiclassical limit of quantum mechanics (Is the Moon there when

somebody looks?)", in R.J. Russell, *et al.*, eds., *Quantum Mechanics: Scientific Perspectives on Divine Action* (University of Notre Dame Press, 1993), p. 52

9. R. Courant and H. Robbins, *What is Mathematics?* (Oxford University Press, 1941) pp. 28, 30

10. D. Zagier, "The first 50 million prime numbers", *The Mathematical Intelligencer* 0 (1977) pp. 16, 9, 7

11. D. Spencer, *Exploring Number Theory With Microcomputers* (Camelot, 1989) p. 35

12. Y. Motohashi, quoted in K. Sabbagh's, *Dr. Riemann's Zeros* (Atlantic, 2002) p. 17

13. R. Padma and H. Gopalkrishna Gadiyar, "Renormalisation and the density of prime pairs" (preprint, 1998) [http://xxx.lanl.gov/abs/hep-th/9806061]

14. from I. Peterson's "Math Trek" column, *Science News* 12th February 2000 [available at http://www.sciencenews.org/20000212/mathtrek.asp, also archived as http://tinyurl.com/d3qqq9s]

15. from http://www.sdsc.edu/tmf/Examples/Rie/rie.html [also archived as http://tinyurl.com/crctpc8]

16. from http://dspace.dial.pipex.com/town/way/po28/maths/formulae.htm [also archived as http://tinyurl.com/cvujhtj]

17. W.D. Smith, "Cruel and unusual behavior of the Riemann zeta function" (preprint, 1998) [see http://empslocal.ex.ac.uk/people/staff/mrwatkin/zeta/cruel.pdf, also available as http://tinyurl.com/2vozykt]

18. M.C. Gutzwiller, *Chaos in Classical and Quantum Mechanics* (Springer, 1990) p. 310

19. This example of a fractal is made up of a set of points in the plane (fractals can exist in more or fewer than two dimensions). The set will continue to display similar intricacy and complexity as we zoom in or out from the rectangular window on the set which the illustration shows, with certain geometric features repeating themselves at different scales (a phenomenon known as *self-similarity*).

20. I. Stewart, "Jumping champions", *Scientific American* 283 (December 2000) p. 80

CHAPTER 40 [pages 201–211]

1. Although author Mark Haddon was careful to maintain an ambiguity throughout the novel as to the exact disorder affecting his main character, some editions of the book have featured a publisher's description on the cover which *did* mention Asperger syndrome. This has led to some controversy, with Asperger syndrome experts claiming that it's not an accurate portrayal, and Haddon maintaining that it's not about any specific disorder.

2. O. Sacks, *The Man Who Mistook His Wife for a Hat* (Summit Books, 1985) p. 191

3. E. Bombieri, "Prime territory: Exploring the infinite landscape at the base of the number system", *The Sciences* 32 (1992) p. 31

4. M. Yamaguchi. "Questionable aspects of Oliver Sacks' (1985) report", *Journal of Autism and Developmental Disorders* 37 (2007) p. 1396

5. D. Tammet, *Born on a Blue Day* (Simon and Schuster, 2006) pp. 1, 8–9

6. Savant syndrome is a rare and poorly understood phenomenon (not recognised as a "disorder" by the medical establishment) which involves a person with some form of developmental disorder exhibiting extraordinary brilliance or ability in a limited field.

7. see p. 125 of Volume 2

8. S.R. Ranganathan, *Ramanujan: The Man and the Mathematician* (Asia Publishing House, 1967) p. 88

9. see `http://empslocal.ex.ac.uk/people/staff/mrwatkin` [also available as `http://tinyurl.com/cckfwv9`]

10. Fundamental physical constants are quantities such as the speed of light, the *fine structure constant* and the *universal gravitational constant*, which are physically measurable rather than deducible from first principles like the familiar mathematical constants π, e or the square root of 2. Some of these (such as the fine structure constant) are defined as ratios, hence are *dimensionless*, meaning that they aren't expressed in terms of units such as metres, seconds or kilograms. Like the mathematical constants, these appear to be "built into the structure of reality".

11. See `http://empslocal.ex.ac.uk/people/staff/mrwatkin/isoc/iradier.pdf` [also available as `http://tinyurl.com/cjgbkey`]. Ayurveda is a traditional Hindu system of medicine.

12. This page has since disappeared, but I believe it was referencing an appendix in J. Satinover's *Cracking The Bible Code* (HarperCollins, 1997)

13. P. Plichta, *God's Secret Formula: Deciphering the Riddle of the Universe and the Prime Number Code* (Element, 1998)

14. self-published via Lulu.com in 2005 (ISBN 9781411658790)

15. `http://nonduality.com/news_archive_october_to_december_2003.htm` [also available as `http://tinyurl.com/ccbzft2`]

16. D. Zagier, "The first 50 million prime numbers", *The Mathematical Intelligencer* 0 (1977) p. 8

17. H. Weyl, *Philosophy of Mathematics and Natural Science* (Princeton University Press, 1949) p. 7

18. E. Bombieri, "Prime territory: Exploring the infinite landscape at the base of the number system", *The Sciences* 32 (1992) p. 36

19. M. du Sautoy, *The Music of the Primes: Why an Unsolved Problem in Mathematics Matters* (HarperCollins, 2003) pp. 84, 313

CHAPTER 41 [pages 213–235]

1. Transpersonal psychology is defined in Chapter 37, Note 3. Jungian psychology could be seen as a subset of this.

2. A. Weil, *Œuvres Scientifiques: 1964–1978* (Springer-Verlag, 1979) p. 173

3. J.F. Burnol, "On Fourier and zeta(s)", *Forum Mathematicum* 16 (2005) p. 836

4. K. Sabbagh, *Dr. Riemann's Zeros* (Farrar, Strauss & Giroux, 2003) p. 248

5. J.E. Littlewood (with B. Bollobás), *Littlewood's Miscellany* (Cambridge University Press, 1986) p. 16

6. review by D. Limi [http://www.villagevoice.com/issues/0316/edlim.php, also available as http://tinyurl.com/bneoms3]

7. from publisher's description of J. Derbyshire, *Prime Obsession* (John Henry Press, 2003)

8. R. Bellman, *A Brief Introduction to Theta Functions* (Holt, Rinehart and Winston, 1961) p. 30

9. G.H. Hardy, *Collected Works*, vol. II (Oxford University Press, 1967) p. 14

10. J. Brian Conrey, "The Riemann Hypothesis", *Notices of AMS* (March 2003) p. 344

11. M. du Sautoy, *The Music of the Primes: Why an Unsolved Problem in Mathematics Matters* (HarperCollins, 2003) p. 22

12. K. Sabbagh, *Dr. Riemann's Zeros* (Farrar, Strauss & Giroux, 2003) p. 89

13. M.C. Gutzwiller, "Quantum chaos", *Scientific American* 266 (January 1992) p. 32

14. M.V. Berry and J.P. Keating, "$H = xp$ and the Riemann zeros", in *Supersymmetry and Trace Formulae: Chaos and Disorder*, ed. I.V. Lerner, *et al.* (Kluwer, 1999) p. 365 and "The Riemann zeros and eigenvalue asymptotics", *SIAM Review* 41 (1999) p. 236

15. M.V. Berry, "Chaos and the semiclassical limit of quantum mechanics (Is the Moon there when somebody looks?)", in R.J. Russell, *et al.*, eds., *Scientific Perspectives on Divine Action* (University of Notre Dame Press, 1993), p. 52

16. T. Gowers, *Mathematics: A Very Short Introduction* (Oxford University Press, 2002) p. 142

17. *The American Heritage Dictionary of the English Language* (Houghton-Mifflin, 1981) p. 1315

18. J. Brown, "Where two worlds meet", *New Scientist* (18 May 1996) p. 30

19. in an interview with P. Hoffman, *Atlantic Monthly* (November 1987) p. 74

20. This concept can be traced back to Rudolf Otto's writing about "the numinous" in *The Idea of the Holy* (translated by John W. Harvey, Oxford University Press, 1923).

21. A. Elon, *Jerusalem: City of Mirrors* (Fontana, 1989) p. 32

22. K. Sabbagh, *Dr. Riemann's Zeros* (Farrar, Strauss & Giroux, 2003) p. 90

23. see http://en.wikipedia.org/wiki/Dirichlet's_theorem_on_arithmetic_progressions

24. Marie-Louise von Franz, *Number and Time: Reflections Leading Toward a Unification of Depth Psychology and Physics* (Northwestern University Press, 1974)

25. There are, of course, many male gardeners, healers and peacemakers, as well as female scientists, architects and business leaders. But suppose we were to present the following list of possible careers to a large group of adolescent males who had been selected by their peers for being decidedly "masculine": businessman, social worker, scientist/engineer, stay-at-home father, policeman, artist, soldier, gardener, counsellor, builder/architect. I think it's quite clear which five would top the list.

26. Although I generally resist using analogies which compare humans to machines, this one is rather helpful, I find.

27. The Qabalistic "Tree of Life" is usually represented as a diagram involving eleven *sephiroth* ("spheres") variously linked by a set of 22 paths (each associated with a letter of the Hebrew alphabet). The sephiroth are all named, each being associated with various aspects of reality and consciousness. Beyond this basic tree, however, tradition dictates that there is a hierarchy of four overlaid trees, each associated with a different "world" or realm.

28. The Higgs boson is an elementary particle first hypothesised in the 1960s and (possibly) first detected in 2012. Its existence (along with the accompanying *Higgs field*) would partially explain how fundamental particles acquire mass and thereby help to fill in some of the remaining gaps in the so-called "Standard Model" of physics. See http://en.wikipedia.org/wiki/Higgs_boson.

29. M. Martin, "Scientist finds new clues to old math mystery" (date unknown) [http://www.weeklyscientist.com/ws/articles/attmath.htm, also available as http://tinyurl.com/ct47k3o]

appendix 15

multiplication represented geometrically

Multiplication can be carried out geometrically like this:

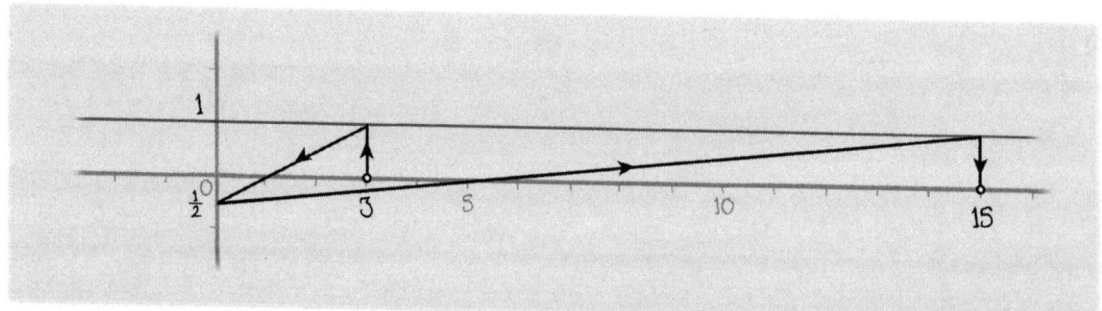

Travel up from 3 to horizontal line of height 1, then travel along a line through 1 until you reach the vertical axis. Finally, travel along a line through 5 until you reach the horizontal line with height 1 and drop back down to the number line.

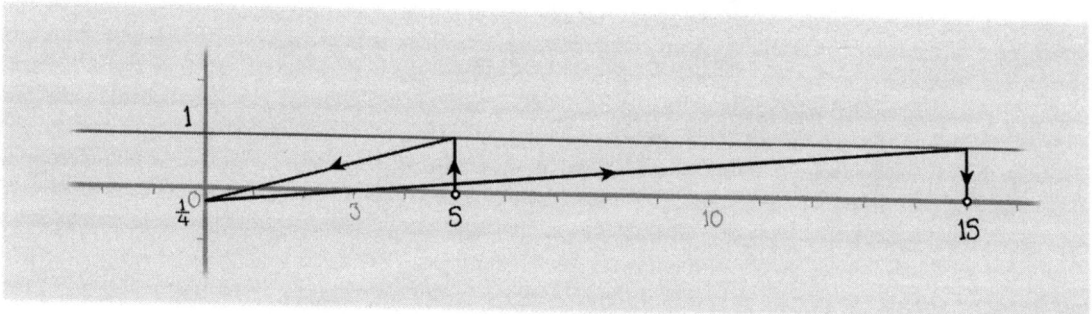

Travel up from 5 to horizontal line of height 1, then travel along a line through 1 until you reach the vertical axis. Finally, travel along a line through 3 until you reach the horizontal line with height 1 and drop back down to the number line.

Here we see geometric representations of 3 multiplied by 5 [and 5 five multiplied by 3], but there's no obvious involvement of "three instances of five elements" [or "five instances of three elements"]. Arguably, though, the construction has been set up so that when you draw the first of the sloping lines, you arrive at a point on the vertical axis which defines a distance from 0 which allows you to divide the vertical unit segment into two [four] parts equal to that distance. So we have a total of three [five] horizontal slabs of equal height:

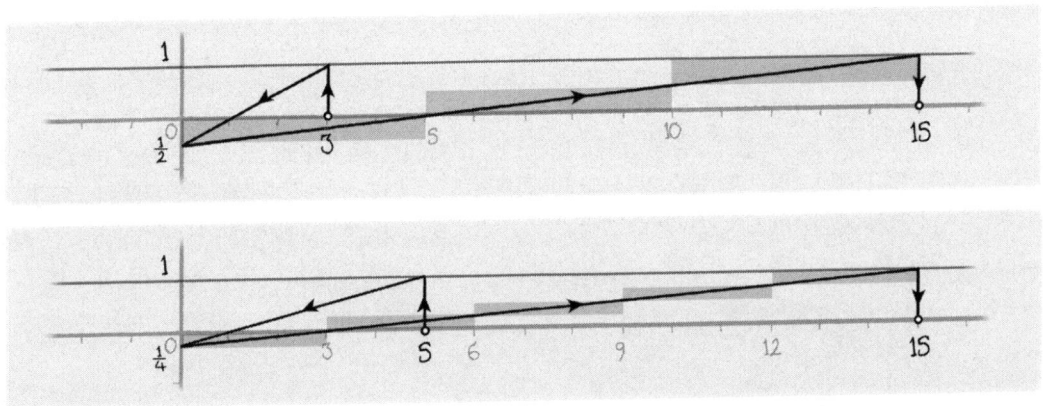

When we draw the second sloping line, we are effectively joining three [five] sloping segments together, each of which covers five [three] units of horizontal distance.

It could be argued that, as pure geometry, there's still no counting necessary – the whole construction exists timelessly and simultaneously, so we can't identify corresponding chunks of the time continuum from this. But, experientially, to construct it and/or comprehend it, there will be some movement of pen or attention (or *something*) across those line segments. We can then divide time up into small chunks containing "corresponding events" in the form of crossings of vertical lines by pen, or in the form of counting vertical strips crossed horizontally.

I suspect that any attempted counterexample put forward would eventually give way to a similar analysis. This argument about the relationship between addition, multiplication and time is something I've come up with myself, not something widely accepted or acknowledged, so I may be wrong, there may be some approach to multiplication which totally contradicts my argument. Still, it's something to think about…

appendix 16

basic Jungian terminology

anima: The archetypal image of woman in the male unconscious psyche, associated with Eros and the "soul". Jung described the anima as "*a natural archetype that satisfactorily sums up all the statements of the unconscious…It is always the a priori element in* [a man's] *moods, reactions, impulses, and whatever else is spontaneous in psychic life.*" The anima is often projected onto an actual woman, initially the mother, later others, and the development of a man's anima reflects the way in which he relates to women. As an archetypal force, the anima takes whatever shape or form is appropriate to compensate for the dominant conscious attitude.

animus: The archetypal image of man in the female unconscious psyche. Personifying "spirit" and "intellect" (Logos), Jung described the animus as "*the deposit, as it were, of all woman's ancestral experiences of man – and not only that, he is also a creative and procreative being, not in the sense of masculine creativity, but in the sense that he brings forth something we might call…the spermatic word.*" He also wrote "*Through the figure of the father he expresses not only conventional opinion but equally what we call 'spirit,' philosophical or religious ideas in particular, or rather the attitude resulting from them. Thus the animus is a psychopomp, a mediator between the conscious and the unconscious and a personification of the latter.*" The animus can be projected onto actual men and the development of a woman's animus reflects the way in which she relates to men.

archetype: The idea of archetypes can arguably be traced back to Plato's notion of "ideal forms", although Jung's concept has several other threads of influence (Burkhardt, Levy-Bruhl, Kant, St. Augustine). An archetype is sometimes described as a "primary structural element of the human psyche", or even an "organ" of the psyche. Jung suggested that they are built in, something we're all born with, the mechanisms whereby our initially chaotic perceptions are gradually transformed into meaningful patterns. They're associated with timeless mythological themes, surfacing in folklore, dreams, fairytales, art, religious ritual and scripture and psychotic hallucinations. Some of the better known examples include the Hero, the Fool, the Tree of Life, the Great Mother, the Divine Child and the Wise Old Man/Woman, as well as the Self and its Shadow. These archetypes supposedly intermingle and are therefore hard to separate (one source of criticism of Jung's theory of archetypes).

collective unconscious: The layer of the human psyche which runs deeper than the personal unconscious, consisting of shared, inherited, archetypal material. Jung wrote *"The collective unconscious – so far as we can say anything about it at all – appears to consist of mythological motifs or primordial images, for which reason the myths of all nations are its real exponents. In fact, the whole of mythology could be taken as a sort of projection of the collective unconscious… We can therefore study the collective unconscious in two ways, either in mythology or in the analysis of the individual."*

compensation: An unconscious self-regulatory mechanism for establishing balance in the psyche. When the contents of the conscious mind become too one-sided (see *functions* below), the unconscious pushes material up into it which "compensates".

functions (thinking/feeling/sensation/intuition): These four primary mental activities occur in opposing pairs: the "rational" thinking-feeling pair and the "irrational" sensation-intuition pair. Usually, one function is dominant in an individual, its opposite being the *inferior function* usually unconscious and often a source of problems. When one half of a pair is neglected, it can result in compensation involving the other half. *"Sensation establishes what is actually present, thinking enables us to recognize its meaning, feeling tells us its value, and intuition points to possibilities as to whence it came and whither it is going in a given situation."*

grail quest: Those parts of the body of medieval stories associated with King Arthur and his knights which concerned the quest for the Holy Grail were of particular interest to Jung, who interpreted them in terms of man's quest to unite with his lost/suppressed/hidden inner feminine.

projection: At automatic process through which contents of the personal unconscious are seen in other people. Jung believed that *"The general psychological reason for projection is always an activated unconscious that seeks expression."* Projection can have the effect of *"isolat[ing] the subject from his environment, since instead of a real relation to it there is now only an illusory one. Projections change the world into the replica of one's own unknown face."* Common forms of this include *anima projection* (a man projecting his anima onto an actual woman) and *animus projection* (vice versa).

psychoid: That which is neither physical nor psychological, but somehow transcending both. Jung believed archetypes to have this "psychoid" quality. The psychoid expresses the unknowable (but somehow experienceable) connection between matter and psyche.

shadow: The archetypal repository for the repressed parts of the human psyche – that which we fear, hate or deny. This can manifest as a fearful figure in dreams or be projected onto others (both personally and culturally). The shadow is meant to be cast by the individualised ego.

unconscious: Those psychic phenomena which lack the quality of consciousness. The unconscious is usually divided into the *personal unconscious* and *collective unconscious*.

unus mundus: "one world" – that which Jung believed to underlie both the physical and the psychological. See *psychoid*.

appendix 17

further thoughts from Barry Jeromson

In response to some of the ideas in Chapter 41, Jeromson has pointed out that many of the quoted male mathematicians who "…metaphorise aspects of mathematics such as prime numbers as a mysterious, elegant, beautiful, etc. woman…lived at a time in Western culture when both sexuality and women were strongly repressed. This repression was very pronounced in fields like mathematics and the mathematical sciences. That being the case, if any projection is going on, is it more likely to be shadow rather than anima?"

He goes on: "Two of the male mathematicians you cite, G.H. Hardy and Alan Turing, were homosexual. How does that blur their comments? How widespread was the use of this metaphor? To my knowledge, nobody ever did any surveys, so this phenomenon might be restricted to a small number of flighty mathematical commentators given to expressing themselves in purple prose.

Someone cited in your chapter raised the point of whether female mathematicians make such comments. Nobody, though, asked the same question in relation to young present-day male mathematicians. If the metaphor…is not common amongst these groups, then this adds weight to my argument that what we see historically is the projection of repressed sexuality – in Jungian terms, projection of shadow.

In describing anything as archetypal, we of course need to define "archetype". If we just mean prototypical, then the natural numbers, as the foundation of all other number systems, are archetypal. However, Jung's archetypes arose in human's most primitive ancestors and reside in the collective unconscious. On the basis of the evidence advanced by Dehaene [see The Number Sense (Oxford University Press, 1997)], therefore, only the first 3 or 4 numbers satisfy Jung's description.

Given Jung's inability to do mathematics (by his own admission) I think it highly unlikely that he would have contributed anything significant to the interpretation of prime numbers, had he lived. Anything mathematical that Jung came up with in later years probably had its genesis in either Wolfgang Pauli or, to a lesser extent, Marie-Louise von Franz."

Index

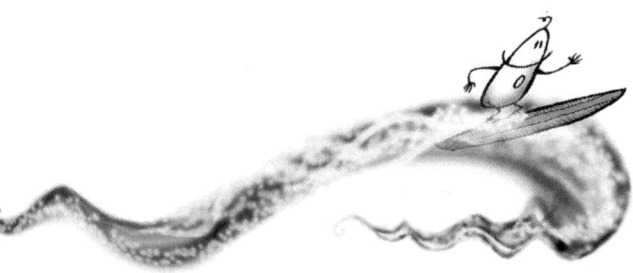

Liberalis is a Latin word which evokes ideas of freedom, liberality, generosity of spirit, dignity, honour, books, the liberal arts education tradition and the work of the Greek grammarian and storyteller Antonius Liberalis. We seek to combine all these interlinked aspects in the books we publish.

We bring classical ways of thinking and learning in touch with traditional storytelling and the latest thinking in terms of educational research and pedagogy in an approach that combines the best of the old with the best of the new.

As classical education publishers, our books are designed to appeal to readers across the globe who are interested in expanding their minds in the quest of knowledge. We cater for primary, secondary and higher education markets, homeschoolers, parents and members of the general public who have a love of ongoing learning.

If you have a proposal that you think would be of interest to Liberalis, submit your inquiry in the first instance via the website: www.liberalisbooks.com.